THE MASSEY LECTURI

The Massey Lectures are co-sponsored by CBC Radio, House of Anansi Press, and Massey College in the University of Toronto. The series was created in honour of the Right Honourable Vincent Massey, former Governor General of Canada, and was inaugurated in 1961 to provide a forum on radio where major contemporary thinkers could address important issues of our time.

This book comprises the 2013 Massey Lectures, "Blood: The Stuff of Life," broadcast in November 2013 as part of CBC Radio's *Ideas* series. The producer of the series was Philip Coulter; the executive producer was Bernie Lucht.

LAWRENCE HILL

Blood: The Stuff of Life is Lawrence Hill's ninth book. His earlier works include the novels *Some Great Thing* and *Any Known Blood,* and the memoir *Black Berry, Sweet Juice: On Being Black and White in Canada.* His novel *The Book of Negroes* won numerous awards including the Commonwealth Writers' Prize, was published around the world, and became a number one national bestseller in Canada. A former journalist with the *Winnipeg Free Press* and the *Globe and Mail,* Hill has travelled widely in Canada, the United States, France, and Spain, and worked as a volunteer with Crossroads International in Niger, Cameroon, and Mali. To encourage the economic and social development of girls and women in Africa, he has supported Crossroads International, currently as an honorary patron, for more than thirty years. He also volunteers with Book Clubs for Inmates and the Black Loyalist Heritage Society of Nova Scotia. Hill lives with his wife, the writer Miranda Hill, and their five children, dividing his time between homes in Hamilton, Ontario, and Woody Point, Newfoundland. He co-wrote the adaptation of *The Book of Negroes* to a six-part TV miniseries and is currently finishing a new novel. For more information on Lawrence Hill, please visit www.lawrencehill.com.

BLOOD

....

The Stuff of Life

LAWRENCE HILL

ANANSI

This edition published in 2013 by
House of Anansi Press Inc.
110 Spadina Avenue, Suite 801
Toronto, ON, M5V 2K4
Tel. 416-363-4343
Fax 416-363-1017
www.houseofanansi.com

Distributed in Canada by
HarperCollins Canada Ltd.
1995 Markham Road
Scarborough, ON, M1B 5M8
Toll free tel. 1-800-387-0117

Distributed in the United States by
Publishers Group West
1700 Fourth Street
Berkeley, CA 94710
Toll free tel. 1-800-788-3123

House of Anansi Press is committed to protecting our natural environment.
As part of our efforts, the interior of this book is printed on paper that contains
100% post-consumer recycled fibres, is acid-free, and is processed chlorine-free.

17 16 15 14 13   1 2 3 4 5

Library and Archives Canada Cataloguing in Publication

Hill, Lawrence, 1957–, author
Blood : the stuff of life / Lawrence Hill.

(CBC Massey lectures series)
Includes bibliographical references and index.
Issued in print and electronic formats.
ISBN 978-1-77089-322-1 (pbk.) (CAN) —.ISBN 978-1-77089-324-5 (html)

1. Blood—Social aspects. I. Title. II. Series: CBC Massey
lecture series

GT498.B55H54 2013 306.4 C2013-903727-6
C2013-903728-4

Library of Congress Control Number: 2013909835
ISBN 978-1-77089-323-8 (US)

Jacket design: Bill Douglas · Text design: Ingrid Paulson
Typesetting: Alysia Shewchuk

Every reasonable effort has been made to trace ownership of copyright materials.
The publisher will gladly rectify any inadvertent errors or omissions in credits in
future editions.

Canada Council   Conseil des Arts   ONTARIO ARTS COUNCIL
for the Arts   du Canada   CONSEIL DES ARTS DE L'ONTARIO

*We acknowledge for their financial support of our publishing program
the Canada Council for the Arts, the Ontario Arts Council, and the Government of
Canada through the Canada Book Fund.*

Printed and bound in Canada

MIX
Paper from
responsible sources
FSC® C004071

ANCIENT FOREST ™
FRIENDLY

For my son, Andrew Raymond Savoie Hill,
who works with abandon
travels with gusto
and with his deliberate and diplomatic hand
writes the most big-hearted Father's Day cards

CONTENTS

"Sometimes I look at people and wonder if they are related to me. I do this in public places and private spaces...I have indulged in this curious pastime since I was eight years old, when I first understood that all but one of my mother's family had become white."
— Shirlee Taylor Haizlip, *The Sweeter the Juice*

*

"There is no expiation except with blood."
— *Sipra*, the ancient Judaic commentary on Leviticus

ONE

GO CAREFUL WITH THAT BLOOD OF MINE: BLOOD COUNTS

ONE SUMMER MORNING, when I was a child, I was on all fours, playing hide-and-seek on a Toronto schoolyard, when my left wrist began to tingle. I looked down and noticed a broken beer bottle. Turning my hand, I saw more blood than seemed right. It was pouring out of me. I stood, let out a cry, crossed the street, and began running. We lived ten houses up the street, less than two hundred metres away. I got ready to shout out for my mother just as soon as she could hear me. Would I have to go to the hospital? How many stitches would it take to impress my friends? This was a deep cut. Lots of blood. Perhaps I would need twenty stitches. Maybe thirty. Three or four wouldn't earn bragging rights. As I ran, I held out my left arm to direct my splashing blood onto every single sidewalk panel, each one just over a metre long. I slowed, when necessary, to ensure that the bright red trail remained unbroken. Later, I wanted to be able to walk with my friends up and down the street and say, "Look! That's my blood!" Once I reached 20 Beveridge Drive, I turned into the driveway, forgot about the trail of blood, and began screaming. By now, I was hyper-ventilating. I terrified my mother when I burst into the house with blood still flowing out of me. She drove me to the hospital.

A few hours later, with three or four measly stitches in my wrist, I was back home. Inspecting the sidewalk proved something of a disappointment. The dramatic red trail had already turned rust-brown. No one would even recognize it as blood, unless I pointed it out and insisted. I told a friend or two, but they were so supremely unim-pressed that I gave up with the story. I did, however, study the splatter every day as I walked up and down the street. My blood clung to the sidewalk for a respectable period of time — a good week or so, until rain washed it away.

Looking back, I wonder about the mad impulse to hold out my arm and splash every sidewalk panel. I wanted to mark the earth with my own sacred fluid. Look here! This is me! This is proof of my very life, here in this long line of bloody splotches on the sidewalk. The blood had appeared so hot, fresh, and significant when it was spilling from me. But hours later, when it had been downgraded to a mud-brown trail, my accident could no longer be heralded as special or sacred, because the trail I had left no longer looked like blood.

In a separate incident, when I was about eight years old, I crashed, arms first, through a glass door at a cottage in southern Ontario. I still have the scars—right wrist, left bicep—to prove it.

I grew up in Toronto, and you could correctly surmise that I was not wise to the ways of cottages and their storm doors. Ours was not a cottage family. We didn't own one, rarely visited them, and in fact I don't recall my parents ever taking my brother, sister, and me to one before. They most certainly didn't do so after. Shall I say, tongue in cheek, that cottaging was not in our blood? My parents were American immigrants. So let's blame this utter lack of Canadianism on them. White mother, from Chicago. Black father, most recently from Washington, D.C. The day after they married, they said goodbye to the United States and moved to Canada, where they figured life would be easier for them, and for my brother, sister, and me. I supposed they hadn't factored glass doors at cottages into the equation. And why should they have? My father was urban, educated, and lower middle class; cottaging seemed to be about the last thing on his radar. Why trade a perfectly serviceable bungalow in suburban Toronto for a smaller, dirtier house with one tiny bathroom, linoleum floors, and busted screens that admitted every manner of mosquito and horsefly? And pay for this supposed luxury, on top of that?

So we never went to cottages. Except this one time, when we visited a small lakeside cottage with another Toronto family—also black, and also with parents who had left the United States to live and to raise children in Canada.

Not long after we arrived, both sets of parents went out for what later seemed an interminable walk in the woods. My older brother and Alan, a boy from the other family, promptly climbed into a boat and rowed into the middle of the lake. Well. Only mildly risky activity, in comparison to what I got into. That left three children in the cottage under the age of ten: my sister, Karen, our friend Sharon, and me. It did not take long to make Karen and Sharon hate me. Boys are gifted in the art of getting girls to hate them. I don't remember what I did to get excommunicated, but soon enough I found myself locked outside a glass door and wanted to get back inside. Where I could infuriate my sister some more. Where the mosquitoes would be less numerous. They were homing in on me, whining like a choir of enemies, saying, "We're going to suck your blood, and we are so many that there's nothing you can do about it." Outside, I would surely lose some blood to the mosquitoes. But I would have lost a whole lot less if I hadn't tried to rectify the situation. I rattled the door frame, but the lock held. While Karen and Sharon enjoyed my plight, I raised my hands and arms and banged on the door. Nothing. I banged once more, and crashed through the glass.

Travelling through glass is not an advisable way to make your point, beat your sister in a contest of wills, or enter a cottage. I don't remember pain, but I do recall terror. I was struck dumb with fear, because as I studied my right wrist and my left bicep — the two parts of me that had been ripped open — I gazed into deep wounds. I saw white inside those cuts. Was it perhaps a ligament? I was not sure exactly what bodily tissue revealed itself to me, but as the blood began to gush, the last thing I wanted to do again was to look deep into my own body. I understood that I had bones, ligaments, muscles, and blood, but to witness them seemed terrifying. Their proper state was invisible, and neatly tucked away under my skin. I slapped my right wrist and left bicep hard against my chest, locking them into place and dreading the hour when someone might insist on prying them apart. The terror that caused me to clamp my arms so firmly against my body was surely a good thing, in that it elevated my arms, applied pressure, and reduced bleeding during the hour or so that it took my parents to return from their walk in the woods.

With my arms still tight against my chest, my father eased me into the back of a Volkswagen Beetle. I believe he sat back there with me while his friend drove us to see a doctor. I don't remember any pain in the doctor's office, either — although he coaxed me to release my chest and open my arms, and must have injected a local anaesthetic into me before he stitched me up. Only when the cuts were closed — I remember twenty-four stitches on the wrist and seventeen on the bicep — was I ready to look down again. There was no more blood. It was safe to look at myself again. Soon I would be able to return home to brag to my friends about my cuts and stitches. But I would never forget the sight of my own bodily tissues, and the bubbling up of my own blood, which seemed, to me, symbols of my own mortality—symbols I was not ready to confront.

I have shared two intimate stories from my own life to underline the different ways that I saw—and that we all perhaps see—blood in the body. In the first instance, I was imagining immortality. Blood splashed down on the sidewalk was a sign of my own life. My own person. As the blood fell from me, I imagined that I was marking the sidewalk forever. I had seen some children scratch their initials into wet concrete, but this was far more dramatic. In the second instance, crashing through a pane of glass actually offered a window into something that I was not supposed to see: my own blood and guts—my own mortality. Seeing deep into my own body made me feel that I was at risk of losing all that was supposed to be kept sealed and locked inside me. So from an early age, I came to read blood in two distinct ways: either as a sign of impermanence or as one of immortality. And as the years have passed, I have learned that in between those two extremes, there are endless ways to imagine the meaning of blood—ours, and that of others.

Almost half a century after crashing through a cottage door, I had my blood taken in a medical lab. Just a routine checkup. A prick in the crook of the arm, and no, I didn't look. On my laboratory requisition, there were boxes corresponding to some fifty possible tests. Liver function. Kidney function. Cholesterol levels. Potassium. Blood glucose. Blood glucose averaged over the last three months. If all these boxes were ticked, and the results published, some of my most intimate details would be made public. The nurse worked silently. Each time she changed a vial, I heard a little *pop*. I counted the pops. One. Two. Three. Only three vials that day — not too much of an ordeal for me. Within a minute, she had extracted the blood and rolled labels onto the vials. What a waste of my blood it would be if someone mislabelled the vials. I would have had my own arm pricked and blood drawn for nothing. You want your blood to count. To the nurse or the lab technician, it's a hazardous substance. But to you, the rightful owner, it reflects your very life. O lab technician, go careful with that blood of mine. Treat it right. And once you have emailed the test results, please shred my papers before you throw them out.

I HAVE HAD A LIFELONG OBSESSION with blood, and I'm not the only one. As both substance and symbol, blood reveals us, divides us, and unites us. We care about blood, because it spills literally and figuratively into every significant corner of our lives.

It's hard to imagine a single person in a school, restaurant, theatre, hockey arena, hospital room, or bookstore who does not have a set of personal stories about blood. Maybe it was the blood of a distant ancestor, persecuted because his or her blood was deemed to be impure. Maybe it was a grandfather who fell under the blade of a farm instrument and bled to death in the fields. Maybe it was an aunt who donated plasma weekly for decades, or a sister who won international attention for designing a more effective way to kill cancerous white blood cells before they multiplied madly and killed the patient. Maybe something happened to you in the blood lab, or in the operating room, and lodged so deeply in your mind that you have passed the story along to every single family member. Blood keeps you alive, for sure. Yet, the very blood in your veins and arteries can suddenly betray you. One day you feel healthy and have just hiked up a mountain with the person you most love in the world, and the next day what you thought was a routine blood test tells you that you have prostate cancer and had better decide, pronto, if you're going to opt for surgery or radiation, or tempt the gods by doing nothing at all. Blood is the lubricant of our bodies and the endlessly circulating river supplying oxygen and nutrients to our cells. But it is far more than a sign of your physical health, or an omen of your mortality. It has the potential to reveal your most hidden secrets: How is your cholesterol level? How much alcohol have you consumed? Have you been snorting cocaine? Are there any other residual traces that might scare off an employer, or lead a life insurance company to deny your application? What has been the average amount of sugar in your blood over the past ninety days? Did you cheat in that Olympic marathon race? Are you the father of that child? Blood won't tell all. But it can tell enough to get you in a whole lot of trouble.

On the flip side of trouble lies salvation. Through blood, many people commune with God. For centuries, humans spilled blood to seek purification, be released from sin, placate the gods and ensure that the sun would rise the next day and that the earth would offer bountiful nourishment.

Blood speaks to our deepest notions of truth and sanctity. Blood can be used in a court of law to vindicate or convict us. It is one of the most sacred gifts a person can offer, but if it is not safe and pure, that same gift can kill not just one person but many who receive the blood products that it helped create. Blood has been employed in the most outrageous ways to divide human beings and justify crimes beyond heinous, and it has the ability to unite us in the most noble ways.

Blood counts in virtually every aspect of our being that matters deeply. If you are fighting for your life, or caught in a downward slide and soon to be facing death, the things that you care about and the things that you hope for after you have departed this earth are likely to be related in some way to questions of blood. That daughter of yours, who has von Willebrand disease: Will she get the right clotting products so that she can give birth to a healthy child and stay healthy herself? The brother who has leukemia: Will they find a suitable bone marrow donor? The sister who is working for Doctors Without Borders in Kenya: Will she have enough anti-retroviral drugs for the people who come to her AIDS clinic? These matters weigh on us. I can't think of another bodily substance that penetrates our hearts and minds so profoundly.

In this book, I will approach blood from five different angles — all of which fascinate me personally and fuel my own obsessions. I will touch from time to time on the physical properties of blood, and on its intersections with medicine, both ancient and modern. Beyond how blood functions in the body, I am interested in how it weighs on the human mind, and how it influences our perception of who we are, to whom we belong, and how we experience our own humanity.

In chapter one, I will explore how our own notions of blood have evolved over thousands of years, share an overview about its nature and functions, address some of the different ways that blood defines men and women, and ponder how our own blood betrays us.

Chapter two will investigate how blood reflects our deepest notions of truth, honesty, and morality. When blood spills, we demand — in literature and in reality — that it exert a downstream effect. There should be consequences, which satisfy us morally. You can't just go spilling blood without cause or effect. Changing the nature or composition of our blood, as well as how we give and receive it, sometimes raises not a single eyebrow, and at other times it incites trenchant judgement. Our response to alterations of blood points to some of our most enduring values, so chapter two will ponder some of the more provocative aspects of changing our blood in medicine and in sport.

A meditation on blood and belonging will form the essence of chapter three. What does our blood tell us about what we are supposed to do, to whom we belong, and the rights that we enjoy or are denied? How do we use blood to differentiate between groups of human beings, and what does that tell us about ourselves?

In chapter four, I will explore the idea that the exercise of power, control, and public spectacle depends on bloodshed. Where you find some people dominating others — to make a revolution stick, or to quash one, or to vilify enemies, or to demonize an entire minority group — you will have no need of hound dogs to find a trail of blood.

The final chapter will delve into how blood offers up our deepest secrets and revelations. Who did we used to be, before we tried to fool the world and acquire an entirely new personality and public identity? Why and how have people slid from one entrenched identity to another? What truths, including the inconvenient ones that can indict us in a court of law or in the court of family judgement, lie in our blood? To whom are we most distantly related?

Blood reveals us and protects us. It's a curse, and it can be a sign. In Exodus, the blood of the lamb protects the Israelites from the avenging Angel of Death sent to kill the first-born sons of all Egyptians, who are responsible for the enslavement of the Jews. By smearing lamb's blood on doorposts, the Hebrews signify their innocence and their homes are passed over.

Blood can also be a gift. At the Last Supper, Jesus tells his disciples that their wine is his blood and instructs them to drink it in memory of him—a practice and a belief that are still part of the Catholic Mass (the Eucharist, for Anglicans; the Holy Communion, for other Protestants). In religious lore, saints have shown stigmata—bleeding hands that mimicked Christ's wounds from being nailed to the cross.

Blood is not just a symbol in religion. It's a symbol in literature. In storytelling, it is integral to the very way we speak and express ourselves. Iambic pentameter, used in much poetry and in Shakespeare's plays, is said to best capture the rhythm of human speech. Its emphasis—an unstressed syllable followed by a stressed one, da-DUM, da-DUM, da-DUM, da-DUM, da-DUM—lodges in the memory and seems familiar on the tongue and to the ear. Perhaps that's because it is also the sound of the human heart. It is the sound of blood coursing through our bodies.

And this is where we find ourselves, when we behold great art. Right in the core of our bodies, deep in the midst of our arteries.

When I was a child, I had the fortune to have a mother, Donna Hill, who read poetry to me at bedtime. She was a kickass civil rights activist who shook her white friends and family to the core and turned her own life upside down when she fell in love with a black graduate student in Washington, D.C., and moved with him to Canada. However, at the age of three, when it was time for bed, I neither knew nor cared about those things. What I cared about was that a gentle, loving soul named "Mom" would summon all of her enthusiasm and pitch it into her nightly poetry readings. My favourite of all was her rendering of the poem "Disobedience," by A. A. Milne, which begins like this:

> James James
> Morrison Morrison
> Weatherby George Dupree
> Took great
> Care of his Mother,
> Though he was only three.

There, beating alongside our pulse, are the playful, absurd, seductive sounds of the early twentieth-century British writer best known for creating Winnie-the-Pooh. Milne entered our imaginations first and foremost through our ears, by mimicking the sounds of our heart. When you read Milne's poetry aloud, it feels as if you are swimming in your own bloodstream.

It is not just poetry that climbs into your body. In jazz and rock 'n' roll, the driving bass beat holds the music together. The bass beat gets you dancing. You want to slide into bed with it. Forget the lyrics. The bass is where you feel the music. Deep down, in your bone marrow and in the pulsing of your blood.

OUR NOTIONS OF BLOOD have evolved over thousands of years, and our understanding of its nature and functions has shaped our ideas of ourselves. Blood acts as a mirror, reflecting the march of life, of ages and civilizations. It speaks of our beliefs and prejudices, of our potential and our limitations as flawed beings. And like all things biological, chemical, and physical — the mystery of nature — it is governed by its own set of rules and regulations. Indeed, the ways that we identify and interpret the biology of blood affect our self-concept, individually and collectively.

Blood has some four thousand components. A drop of blood the size of a pinhead is teeming with quantities of cells that seem unfathomable: 250 million red blood cells, 16 million platelets, and 375,000 white blood cells. If you can imagine blood in a test tube, separated by means of a centrifuge into its key parts, you will notice three distinct substances, each with its own colour and function. Let me quickly mention them, from top to bottom in their separated forms in our imaginary test tube.

Plasma comes on top, as it is lighter than the other key blood ingredients. Made up of about 92 percent water (water accounts for about 50 percent of our blood), it is a yellowish, straw-coloured alkaline fluid. In it, you find many dissolved solids such as glucose (or sugar); proteins such as albumin, which controls the flow of water in and out of the bloodstream; hormones such as erythropoietin; and insulin, salts, lipids, and waste products such as bicarbonate ions, amino acids, and blood cells. I liken plasma to a river, offering a delivery system for the ingredients in blood, as well as carrying products that help regulate bleeding and clotting.

It is possible to donate and receive plasma separately from other blood products. When blood is withdrawn from the body, the red blood cells, white blood cells, and platelets are separated and returned to the body of the donor, minus the plasma. Some of its key medical uses are to help people cope with bleeding or clotting disorders, recover from burns, deal with immune deficiencies, and survive complications resulting from bone marrow or organ transplants. Plasma can be stored for longer than regular blood products, and it can be frozen or dried for easy transportation. One additional advantage to plasma is that the donor's body can replace plasma much faster than whole blood.

In the tube of blood whose parts have been spun and separated in a centrifuge, the middle of the three layers contains the white blood cells and platelets. White blood cells are pale in colour. They are also known as leukocytes or white corpuscles. They come in different varieties. The primary roles of the white blood cells are to remove waste from the blood and to fight against infection. The language used to describe white blood cells is strangely military. They are said to surround and devour bacteria. They engulf, digest, and destroy invading micro-organisms. One type of white blood cells — accounting for about one-third of a healthy person's white blood cell count — is known as lymphocytes. These include helper cells, suppressor cells, and natural killer cells. The killer cells are labelled "natural" after their function, which is not to attack invading organisms but to destroy the body's own cells that are cancerous or carrying viruses.

The white blood cells are commonly likened in our language to soldiers going to war on behalf of the nations that are our bodies, identifying, targeting, and destroying foreign invaders. The war and battle metaphors we employ — influenced by the writings of Louis Pasteur in the 1800s and reinforced by U.S. president Richard Nixon, who in 1971 signed the National Cancer Act and declared a "war on cancer" — offer one way to contemplate human biology. They certainly provide us with a method to imagine the body's efforts to deal with disease and infection. At the same time, they are at risk of leaving us with the impression that people who succumb to illness simply did not try hard enough, and that people who overcome the same illnesses are stronger, more courageous, or have more valour. It is a striking way to refer to our own bodily processes, but there you have it.

Platelets are, with white blood cells, part of the thin middle layer separating the plasma from the red blood cells. Platelets are fragments of blood cells called megakaryocytes, which reside in the bone marrow. (All blood cell lines, including platelets and red and white cells, originate in the bone marrow.) Platelets live for a short time — only a week or so. The human body produces about one hundred trillion new platelets every day. Their function is to aid in recovery from injury. If something pierces a blood vessel, platelets stick to the damaged lining and clump together. This process aids in coagulation — a blood-thickening process that stops the body from bleeding. The clotting process begins within seconds of an injury. Standing on guard and ready to self-correct, your blood organizes itself to prevent a hemorrhage. Otherwise, copious amounts of blood could drain out of you. A simple puncture of the body, left unattended, could be fatal. You would be like a bicycle inner tube when the tire rolls over a nail, with no patching gear within reach. But your body has its own patching kit. It knows how to clot. A clot can be fatal in the wrong place and for the wrong reason: say, if it is travelling toward your lungs or brain. Then it is known as an embolism. But you want the clotting function to work perfectly and immediately when you nick yourself with a kitchen knife. To me, the platelet is the nurse or doctor in your veins, ever ready to sew you up when you have been shot.

The bottom part of the imaginary tube of blood consists of red blood cells, also known as red corpuscles or erythrocytes. Round in shape and slightly concave on each side, they are the most numerous cells in the blood. Some five billion of them exist in one millilitre of blood. I think of the red blood cell as the cell of love. In contrast to the soldiering white blood cell, and the platelet with its emergency room services, the red blood cell is your bedmate. It is all about giving. The red blood cell lives for only 120 days, but what an ardent lover it is! You should salute your white blood cells and thank your platelets, but the red blood cell deserves your love. It kisses your cells with the gift of oxygen, and it is a non-stop kisser. Your body produces millions of red blood cells every second.

In humans, the blood is red thanks to iron and hemoglobin, an oxygen-carrying protein in the red blood cells. In its appearance, blood stands alone and virtually unmistakable. How often are you wrong when you think you see blood? Thanks to the presence of iron, which is also responsible for the rusty-red beaches of Prince Edward Island, blood is bright red. No other bodily fluid or tissue resembles it. I don't often see the colour of arterial blood in nature. The closest I have seen to blood-red is a sunlit field of poppies. The sight of flowering poppies arrests me, every time. I have to stop and stare at it. I take in a breath, and never fail to think that the field before my eyes is beautiful. Silent under the skies, teased by the wind, it resembles a vast blanket of undulating blood.

But even the colour of blood varies, slightly, and has led many people to wonder if it is blue when not exposed to oxygen. Blood is never blue in human beings, but given the way that light can strike fair-coloured skin, it can sometimes appear that way from outside the body. Indeed, the term *blueblood*, which means a person of noble ancestry, derives from the idea that the venous blood may seem to have a blue tint through the light skin of a person freed from the burden of having to work in the sun. Indeed, polo — a game of the very rich — is sometimes described as a blue-blooded sport. The "colour" of our blood is just one example of how we have uniquely attached meaning and metaphor to blood as a way of differentiating ourselves from others, in this particular instance as a marker of class superiority. Just as quickly as blood can elevate your status, it can denigrate you. A "bloody fool" is an idiot—perhaps dirty, and possibly blood-spattered. The "bloodthirsty masses" are the last thing from genteel. On the contrary, they have empty bellies and, lacking food, insist on violence.

Much as some people have found it convenient and reassuring to imagine that their blood is so special that it acquires a different colour, blood in any human body is bright red when freshly oxygenated and travelling via the arteries to deliver oxygen to the body's tissues. But it is a darker red in the veins, when it is on its way back to the heart for another infusion of oxygen.

When I think of hemoglobin, I imagine millions of miniature versions of Sisyphus. As a punishment for deceitfulness, Sisyphus, a king in Greek mythology, is sentenced to the interminable task of hauling a boulder up a hill, only for it to roll down again so that he has to push it right back up. Unlike Sisyphus, hemoglobin isn't always struggling against gravity. For hemoglobin, the struggle is the laps that it must run around the body— laps that accelerate as the body works harder. The endless task of hemoglobin is to bind itself to oxygen and haul the oxygen to tissues throughout the body.

It takes blood about a minute to circulate through the resting body. When you get to work—chopping logs, hauling laundry, chasing toddlers, or trying to win a dragon boat race — you oblige your blood to work harder. As your arms and legs speed up, your blood is like a stagehand, supplying props at a furious pace as the play unfolds.

The best endurance athletes — especially in the ultimate cardiovascular tests, such as running 42.2 kilometres or racing a bike for three weeks through both the Alps and the Pyrenees — are the ones who transfer oxygen most effectively from their red blood cells to cells in their muscles. After refuelling in the lungs and being pumped back out by the heart, hemoglobin, in its oxygen-rich state, is called oxyhemoglobin. But once it unloads the oxygen at its destination points, it becomes hemoglobin again and scrambles through the veins back toward the lungs for another hit of oxygen, only to recommence its endless trucking route. Pity the hemoglobin of an elite runner in the Boston Marathon, or of a cyclist in the Tour de France. All work and no glory. No wonder Lance Armstrong and a legion of other cyclists opted for blood doping, withdrawing and later re-transfusing their own blood to deliver oxygen more effectively to their overworked muscles.

An average adult has about five litres of blood, representing about 7 percent of their body weight. Blood, like just a few other body parts, such as hair, fingernails, and toenails, is always replenishing itself. Dying and growing back. And like hair and nails, blood will regenerate if you lose it intentionally or accidentally. The bone marrow constantly produces red blood cells, white blood cells, and platelets. And the body can replace donated plasma within hours. But beware: not too many body parts work this way. Lose 'em once, and—like an arm or a leg—they're gone forever. But you can afford to lose up to about 40 percent of your blood and still survive, if you don't lose it too quickly.

There are various types of blood. One common grouping is A, B, AB, and O. And then you can be Rhesus (Rh)-positive or Rh-negative. You have a certain blood type, and that blood type matters. Most blood types cannot be mixed with other blood types. If they are, the person receiving mismatched blood could die.

It is humbling to contemplate all that blood does. In addition to lugging oxygen to all bodily tissues, blood delivers nutrients such as amino acids, fatty acids, and glucose. It carries away waste such as carbon dioxide and lactic acid, detects and attacks foreign invaders, coagulates to stop bleeding, regulates body temperature, transports hormones, detects tissue damage, and is responsible for hydraulic functions—an oddly formal term for the task of blood during sexual activity. I imagine blood as a happy workaholic, humming away and in a state of constant calisthenics as it nourishes us, lifts us into arousal, does battle with invaders, and replenishes itself.

Blood, like human identity, is ever shifting. Just when it has acquired a certain personality or chemical balance, it gets thrown into turmoil and must regain equilibrium.

It can manage stress, but only so much. I imagine blood as the planet earth, and all of the bodily reactions to disturbances — be they sugar or alcohol — as gravitational forces, pulling everything back to the ground. You're in trouble when your gravitational forces don't work, or when you've overdone it so much that your body just can't cope. That, by my way of thinking, is like shooting a rocket through space with no way of bringing it back down to earth.

Travelling like an indefatigable river along its intertwined circulatory systems, passing through the heart and lungs and feeding the rest of the body, blood keeps us alive and has forever held us in its debt. To it we owe our daily health. To it we pay ransom — insulin injections, chemotherapy treatments, bone marrow transplants, the use of clotting factors — when order turns to disorder in our arteries and veins. And from it we build a frame to envisage our own humanity. We let it run from our veins as a gift to others in failing health. And if we believe in a superior being, we give it up as an offering so that we may go on living.

WHAT WE NOW KNOW about blood seems all the more astounding when we think about where we have come from. For some two thousand years, philosophers and physicians imagined blood as one of the fundamental characteristics of our body and soul. We linked it to the spring, the air, and the liver.

Thanks to the theories of figures such as Hippocrates,

in 460 BCE, and Galen, in about 200 CE, we came to believe that sickness arose as a result of disequilibrium between four key parts (or humours) of the body: blood, yellow bile, black bile, and phlegm. Hippocrates inspired the Hippocratic oath and is often referred to as the father of modern medicine. Claudius Galen proved the presence of blood in the arteries, and argued that arteries and veins are distinct and that the liver has a key role in blood production. "The liver is the source of the veins and the principal instrument of sanguification," Galen wrote in *On the Usefulness of the Parts of the Body.*

Galen argued that the preponderance of one particular humour went so far as to determine a person's basic personality type. One might be sanguine, choleric, bilious, or melancholic—words and concepts that continue to resonate with us today. Blood, for example, was said to quicken the spirit, and the adjective "sanguine" derives from the old French word *sanguin* and from the Latin *sanguineus* (meaning "of blood"). It refers to a person who is courageous, loving, and optimistic, especially in difficult situations. The *New Oxford American Dictionary* offers the following definition of "sanguine" in the context of medieval science and medicine: "of or having the constitution associated with the predominance of blood among the bodily humours, supposedly marked by a ruddy complexion and an optimistic disposition."

But too much of any humour would create a dangerous disequilibrium in both temperament and health— the elusive "mind-body" balance we still search and long for today. Traditional Islamic medicine and the Ayurveda medicine of ancient India suggest food and diet as one means to correct imbalances of the humours.

Another was the technique of bloodletting, or phlebotomy. In retrospect, it is sobering to imagine how many thousands of patients have died from bloodletting or its complications. I, for one, hate having my blood played with or withdrawn and feel grateful in the extreme for Louis Pasteur and Robert Koch, both nineteenth-century scientists who demonstrated that inflammation results from infection, thus obviating any need for bloodletting.

Bloodletting is still practised in a few ways. We donate blood, have it withdrawn for laboratory tests, and use it to treat problems such as polycythaemia (an abnormally high concentration of hemoglobin in the blood) and hemochromatosis (a hereditary disorder in which excess iron is absorbed through the gut and deposited in tissues).

For thousands of years, physicians have used leeches as a bloodletting device. They, like some other animals, such as mosquitoes, lampreys, and vampire bats, have figured out that sucking other animals' blood is an effective shortcut to a rich, nutritious meal. I wonder who, in medical cultures in ancient Egypt, Greece, and India, came up with the bright idea of ushering leeches onto human skin for the purposes of bloodletting. Someone must have stepped back and muttered, "But there must be a use for this worm that annoys me so."

Leeches still have a role in modern medicine, particularly in reconstructive or plastic surgery. They dilate the blood vessels and prevent the blood from clotting, and are especially useful after surgery in promoting the flow of venous blood. They have proved useful in the reattachment of body parts such as fingers, hands, toes, ears, noses, and nipples. Because veins have thin walls, they can be hard to stitch together in surgery. Until the body figures out how to do so again, leeches secrete an enzyme that helps move the blood into the thin and sometimes damaged veins of reattached body parts. They are energetic little devils. A leech can suck more than three times its body weight in blood.

It may be troublesome to imagine a leech — which is basically a bloodsucking worm — attaching itself to your body. But other forms of traditional phlebotomy jump out as being far more invasive, and potentially lethal. I would take a leech over a human bloodletter, any day! Clearly, others feel the same way. Eric M. Meslin, associate dean for bioethics at the Indiana University School of Medicine, told me that while he was visiting the Spice Bazaar in Istanbul in April 2013, he came across a vendor who conducted a brisk business selling leeches. Identified on his storefront as "Prof Dr. Suluk," the man sold leeches for purposes such as migraines, cellulite, low back pain, eczema, and hemorrhoids.

In 400 BCE, the Greek historian Herodotus recommended cupping (the use of a partial vacuum to draw blood) as a means to promote appetite, digestion, and menstrual flow, and to resolve problems such as head-aches and fainting. If blood is removed from behind the ears, he said, it brings about a natural repose. Other spots from which blood has traditionally been let include the knees and elbows. Bloodletting was certainly not limited to one cultural, religious, or geographic group. In addition to the Greeks and the Romans, bloodletting was carried out in Islamic cultures (for example, the Arab queen Zenobia killed King Jothima Al Abrash in this manner). Hindus practised it too.

Bloodletting also entered into ancient Jewish traditions. As Fred Rosner wrote in 1986 in an article for the *Bulletin of the New York Academy of Medicine*, in the third to the fifth centuries CE, the Sages of the Babylonian Talmud held that a learned man should not live in a town that had no bloodletter. Bloodletting was recommended for headaches and plethora (an excess of blood).

The medieval scholar and rabbi Maimonides wrote about the benefits and hazards of bloodletting, but not all Jewish writers believed in the practice. The Old Testament contains, in Leviticus, a prohibition against cutting into the skin. Maimonides said that before bloodletting, a patient should recite a supplication to God for healing, and that after treatment concluded that patient should say, "Blessed art Thou, Healer of the Living."

Over the years, many famous people have died of bloodletting. Charles II, king of England, Scotland, and Ireland from 1660 to 1685, should have been inspired by his own family history to pay close attention to the safeguarding of his own blood. After all, his own father, Charles I, was beheaded in 1649 on the charge of treason. Some of the king's followers dipped their handkerchiefs in his blood. Oliver Cromwell, the revolutionary leader, permitted the king's head to be sewn back onto his body so that his family could mourn properly after the execution. Nonetheless, some thirty-six years later, his son Charles II found himself ill at Whitehall Palace in London. Known as "the merry king" for his philandering, Charles II took to his bed one night with a sore foot. The next day, a barber shaved his head and the bloodletting began. In addition to enduring purging, mustard plasters, red-hot irons, and enemas of rock salt and syrup, Charles II had twenty-four ounces of blood withdrawn from his arms. He suffered a seizure and died.

Napoleon survived a bloodletting and is known to have described medicine as "the science of murderers." Mozart is thought to have died of shock from severe bloodletting, and George Washington lost his life a day after more than 2.3 litres of blood were taken from him, purportedly to help him cope with a cold and hoarseness.

It would be easy to mock bloodletting as pseudo-medicine that hurt or killed thousands of people over thousands of years, and whose widespread use has come to a halt only in the past century or so. But that would be an easy target. It is not hard to imagine the peals of laughter and squeals of disbelief that people might share in one hundred years when they analyze today's medical practices. They will surely shake their heads and say, "What were they thinking?" We will always be in a state of evolution with regard to the perceptions about how our bodies work and how they can be cured of illness and disease.

To me, the interesting thing about bloodletting is how thoroughly it was interwoven with our belief systems, and how long it endured. For two millennia, we coasted with the unassailable idea that a healthy body and a healthy mind should not be burdened by too much blood. In our minds, we removed its impurities to improve our physicality, along with our moods and emotions, by letting our blood run. And ever since, we have been obsessed with the idea of balancing our minds and bodies and improving the composition of our blood. For thousands of years, to spill our own blood in the name of medicine, we used every manner of knife, quill, tooth, lancet, and scalpel. We could have filled rivers and lakes with all the blood we have voluntarily spilled. We did so because of our belief systems. We let our blood run because we had imaginations.

We are always looking for ways to distinguish ourselves from other animals. Let me add one more point of comparison. Can you think of any other animal that cuts itself, or others, to satisfy the cravings of its soul? Other animals will attack if they need food, or run to avoid being eaten, but generally they have the good sense to leave their own blood alone.

THE SEVENTEENTH-CENTURY British anatomist William Harvey—physician to King Charles I—refuted thousands of years of medical thinking when he proved that blood circulates in the body and is pumped by the heart. He dissected live animals to establish his theory. It seems barbaric today, but Harvey had no other means at his disposal to advance his theories. As Thomas Wright notes in his book *Circulation: William Harvey's Revolutionary Idea*, in the early seventeenth century, "Men could no more see blood coursing around their arteries and veins, going to and from the heart, than they could perceive that the earth was spinning round."

In 1628, Harvey confronted and shocked his doubters at the University of Altdorf in Nuremberg. Dressed in a white gown and his head covered with a white bonnet, the diminutive physician instructed porters to affix a live dog to a dissection table, immobilizing it and tying its jaws shut to prevent barking. He plunged a knife into the animal's thorax, exposed its heart, and indicated the rising and falling of the organ. When the dog's heart was in contraction, Harvey severed an artery. The blood spewed forth, showering the closest spectators, several feet away. Thus we finally learned the basics of blood circulation. Understanding this concept opened up the long and painful path toward blood transfusions.

In the mid-1600s, doctors in France and England competed madly for the honour of carrying out the first blood transfusions. Dogs were transfused with the blood of other dogs, and eventually humans were transfused with the blood of calves and lambs. While some of these procedures did not lead to fatalities (possibly because little or no animal blood actually managed to enter the human bloodstream) and were deemed a success, others did result in death.

One of the earliest documented transfusion attempts involved a French physician named Jean-Baptiste Denis, who grabbed a man named Antoine Mauroy off the streets of Paris in 1667 and attempted to calm his agitated mind by forcefully transfusing the blood of a calf into his veins. The physician believed that the mildness and freshness of the gentle animal would, by entering the patient's bloodstream, affect his personality. We now know that a human being is likely to have a severe or fatal reaction to the blood of any animal, because the human body rejects the foreign substance. Mauroy died after a few attempts. His wife brought a complaint to the authorities. Denis and his colleagues countered that the patient's wife had poisoned her husband. A French court eventually exonerated the doctor and charged the victim's wife with murder. She disappeared from the records, and it is likely that she was executed.

The French court finally decreed that no further transfusions were to be carried out without the consent of the French Faculty of Medicine. Shortly thereafter, both France and England banned human transfusions outright.

James Blundell, the nineteenth-century English obstetrician, carried out the first successful human-to-human blood transfusions. In 1818, he used a syringe to extract four ounces of blood from a husband and transfuse it into his wife, to treat postpartum hemor-rhage. She survived, and Blundell went on to carry out another ten transfusions between 1825 and 1830, five of which proved beneficial.

Many patients died in early human-to-human blood transfusions. Looking back, we now know that many such deaths occurred because of mismatching blood types between donor and recipient.

The path toward safe blood transfusions became much more promising in 1901, when the Austrian Karl Landsteiner discovered three blood groups: A, B, and C (later called O). The very next year, Landsteiner's col-leagues identified a fourth blood group: AB. The blood groups A, B, and AB are incompatible with each other. The blood type O, however, is compatible with the others. Within half a decade of Landsteiner's discov-ery, Reuben Ottenberg at Mount Sinai Hospital in New York performed the first transfusion by matching blood types. He went on to perform more than a hundred other transfusions without the problems that had resulted pre-viously from mixing incompatible types.

Initially, blood transfusions required the volunteer donor to be placed next to the recipient so that their veins could be hooked up together. This was known as "blood on the hoof." You can imagine how impractical it must have been to always require donors and recipients to be together, but people were amazingly creative in address-ing the challenge. For example, in 1921, under the super-vision of a man named Percy Lane Oliver, the Red Cross

in London, England, set up a list of people who promised to be available, at any hour of the day or night, when
donors were needed. These volunteers underwent medical exams ahead of time, including tests for blood type
and syphilis. Their telephone numbers were recorded.
At its height of activity in the 1930s, people on this list
responded to nine thousand calls a year. Although the
process of lining up donor with recipient must have been
cumbersome, the emotional connections between the
two people surely heightened the understanding of the
value of the gift for both parties.

But blood on the hoof diminished as a medical necessity as the science of blood storage moved forward.
Sodium citrate was identified as a means to prevent
stored blood from clotting; refrigeration was discovered as a safe means to prolong the shelf life of blood;
and the looming tragedy of war drove us to discover the
possibilities of blood banks so that massive amounts of
blood could be moved to the front lines to save the lives
of injured soldiers. The first blood depot was used in
World War I, in 1917, when the U.S. army doctor Oswald
Roberston used a citrate-glucose solution to store type
O blood, to be used for British soldiers returning injured
after fighting the Germans in the Battle of Cambrai in
France. Use of the citrate-glucose solution made it possible to store the blood safely for a few weeks.

The Canadian surgeon Norman Bethune entered
the picture during the Spanish Civil War. On the side
of the leftist Republicans, Bethune established a mobile

blood service in Spain in 1936. Following the example of a Spanish hematologist working in Barcelona, Bethune travelled to Madrid and set up the service, which brought blood in bottles to Republican soldiers who had been injured at the front. Using a kerosene-powered refrigerator and sterilizing equipment, Bethune's Canadian Blood Transfusion Service was soon making use of 4,000 donors, 100 staff members, and five trucks to deliver blood for 100 transfusions a day. Essentially, Bethune created an early model of the Mobile Army Surgical Hospital (or MASH, for short) units that were later employed by the American army during the Korean War in the early 1950s.

In 1940, the African-American physician Charles Drew — who had studied medicine at McGill University — responded to a blood shortage in Britain during the Second World War. The United States had not yet entered the war, but working for the Presbyterian Hospital in New York, Drew devised a safe system to process, test, store, and ship plasma overseas to Britain. (I will discuss Drew in more detail in the next chapter.) After he directed the Blood for Britain program, the United States entered the war and had to see to its own blood supply needs. The American Red Cross organized a civilian blood donor service and collected some thirteen million units of blood over the course of the war.

The science of transfusion continued to advance, thanks in part to new breakthroughs such as the discovery of the Rhesus blood system (more on this later), the use of better anticoagulants to enhance blood storage, and the use of plastic bags for blood collection. Today, some 92 million blood donations are collected each year worldwide, many of them to be used for transfusions. As a noble and selfless gift, blood is in an entirely different class than money. Dollars from your bank account can be directed to the charity or person of your choice. And you expect, usually, to be thanked and recognized for a cash gift. If you give enough money, you may even get a bridge or building named after you. But for the most part, you have no idea who will receive your blood. You give it on principle. You and the recipient will never meet.

It is not surprising that the evolutionary turning points in the science of transfusion culminated around the major wars over the course of the twentieth century. Blood donations became the ultimate symbol of the gift of life, from civilians to soldiers. Individuals helped save individuals, but also sought to protect an entire generation. But as we will see in the next chapter, blood donation has also come to reflect our deepest discriminations and prejudices, when philanthropy collides with politics and our very worst selves overcome our best selves.

WHILE BLOOD IS UNIVERSAL in its nature and functions, it is also a marker of the gender difference between men and women. I'm sure that I was in no way unique when, at the age of thirteen or so, I walked into the bathroom in our Toronto home and gasped to see blood in the toilet. I knew that my sister, younger by one year, had just been in the same bathroom, and I ran to tell my mother that Karen was bleeding. I was astonished, and intrigued, by the firmness and quickness with which my mother flushed the toilet, told me not to worry, said everything was under control, and dismissed me. Earlier, she had told me about women's menstrual cycles, and later, I believe, she repeated the message. But in the moment, her job was to get me out of the way and show me that what I thought was a major event deserved no attention whatsoever.

Typically, men associate the spilling or the sight of blood as the by-product of accident, sport, or war. Blood, for men, is often romanticized and likened to ritual, honour, and a sign of one's masculinity and courage. When a woman bleeds during her monthly cycle, it is a symbol of coming of age, of fertility, a sign of her sex. We honour people who spill blood in defending their cause or their nation, yet since time began we have found the most curious, inventive, and offensive ways to vilify women for shedding their monthly blood — the same blood that they have needed before or will need again to build the very beginning of a nest inside the body for the earliest promise of human life.

Men in ancient Greece, and surely beforehand, seemed befuddled by women's blood. They knew that they — men — would be in serious trouble indeed if they bled like women did. If men bled for days on end, it was surely the result of illness or injury and they would likely perish. So how could women bleed so profusely and regularly and still not die? This apparent imperviousness to mortal weakness might have been taken as a sign of women's power, or of magical qualities barred to men, or of some sort of gender superiority. But for the most part, men found a much more self-aggrandizing theory: monthly bleeding was proof of women's inferiority.

In the fourth century BCE, the Greek philosopher Aristotle noted in his treatise *On the Generation of Animals* that men's blood was superior to that of women. In the human body, Aristotle wrote, heat transformed nourishment into blood. Males, who had sufficient heat, were then able to embark on an additional step: they could "concoct" (or transform) the blood into semen. On the other hand, Aristotle claimed, women were colder and thus lacked sufficient heat to produce semen. Aristotle's exact words were that "the woman is as it were an impotent male, for it is through a certain incapacity that the female is female, being incapable of concocting the nutriment in its last stage into semen in women." Because women lacked this ability, Aristotle said, they ended up with extra blood in their blood vessels and had to expel it during menstruation. Aristotle's meditations about the blood of men and women helped entrench notions of male superiority and female inferiority that have lasted more than two thousand years.

We can't blame sexism on Aristotle alone, nor can we suggest that he was the first to obsess about women and blood. In *The Curse: A Cultural History of Menstruation,*

Janice Delaney lists a litany of other theories by classical Greek and Roman male philosophers about the fundamental problems associated with the blood of women. In the fifth century BCE, Empedocles, best known for theorizing about earth, water, air, and fire, said women evacuate blood because their flesh isn't as dense as the male's. In the same century, Parmenides — founder of a school of philosophy and author of the poem "On Nature" — said women are hotter than men (this in opposition to Aristotle's later theory about women being cold) and thus produce an excess of blood, but that they gradually get colder until they reach menopause. Galen, the celebrated Greek physician and surgeon, theorized in the second century BCE that women menstruate because they are idle, live continually at home, and are not used to hard labour or exposure to the sun. Other theorists in the eleventh and seventeenth centuries CE speculated that through menstruation, blood escapes from the weakest point of the woman's body, essentially describing the womb — as Delaney notes — as a "defective barrel."

I so wish that these classical philosophers had been around today. Imagine the expression on Aristotle's face if he were disinterred, reanimated, and required to travel — let's send him by Greyhound bus — to Marymount Manhattan College. Why? In June of 2013 at that august institution, he would have been exposed to "Red Howl Moon" — dubbed as the world's first menstrual poetry slam. Today, many people will appreciate the organizers' intentions to, as they say, "bring down the red tent of shame."

Alas, the biological differences are unlikely to change, and for me, one of the most fascinating things about it is how thoroughly clued out, for the most part, men are with regard to the monthly cycles of women. It wasn't until I started researching this book that I learned of the psychologist Martha McClintock's 1971 paper for the journal *Nature*, in which she put forward her theory of menstrual synchrony — that women who live together tend to menstruate at the same time. When I asked my wife and daughters about this later, they replied, "Larry, how could you not know this?" and went on to say that my own ignorance was typically male.

WHILE THE CLASSICAL PHILOSOPHERS used menstruation as a way of calcifying the idea that women are the "weaker sex," religion has further advanced patriarchal notions by focusing on the notion that women's monthly bleedings indicate a lack of cleanliness or purity. In a 2008 issue of the *Internet Journal of World Health and Societal Politics*, co-writers Mark A. Guterman, Payal Mehta, and Margaret S. Gibbs argue that the world's five major religions — Judaism, Christianity, Islam, Hinduism, and Buddhism — all place restrictions on menstruating women. The article, "Menstrual Taboos Among Major Religions," notes that each religion makes statements about menstruation and its negative effect on women, leading to prohibitions on physical intimacy, cooking, and attending places of worship, and sometimes requiring women who are having their periods to live separately from men.

In the Old Testament, in Leviticus 18:19, we find the following prohibition about menstruation: "You shall not approach a woman in her time of unclean separation, to uncover her nakedness." This rule, which influences Christian thinking as well, underpins the "Laws of Family Purity" in the traditional Jewish code of law. These laws forbid physical contact between males and females during the days of menstruation and for a week thereafter. Physical contact means passing objects between each other, sharing a bed, sitting together on the same cushion, eating directly from the wife's leftovers, smelling her perfume, gazing upon her clothing, or listening to her sing. Although these ancient rules have contributed to contemporary prejudices and negative assumptions about menstruation, it must be said that Conservative Judaism has modified the laws of menstrual purity, and Reform Judaism has abolished them as irrelevant, archaic, and offensive to women.

The article's authors note that most Christian denominations do not follow rules about menstruation, but that Western civilization—which is predominantly Christian—has a history of menstrual taboos. Menstruating women have long been believed to be dangerous. In 1878, the *British Medical Journal* postulated that a menstruating woman would cause bacon to putrefy. How that was determined, I would like to know.

Bacon has always done a pretty good job of putrefying on its own — particularly before the advent of refrigeration technologies.

Two vivid examples of the isolation of menstruating women were circulating online recently. In June 2013, the *New York Times* reported that the old Hindu tradition of *chaupadi* (requiring a woman to leave home to stay in a shed, stable, or cave while she is having her period) is alive and well in western Nepal. In addition, in April of the same year, female and male Cambridge University students in the U.K. took to the streets in a campaign called "I Need Feminism Because..." One of the students had herself photographed carrying a poster that obscured her face, and said: "I need feminism because my family wouldn't let me attend my grandad's memorial because MY PERIOD made me UNCLEAN."

Traditions and superstitions around bleeding women have seeped deeply into our culture, from myths surrounding the death of plants touched by a menstruating woman to sailors' stories about the dangers of having a "bleeding wench" on board. They are behind one of the most common slang phrases for menstruation: the Curse.

Still today, the notion of the Curse is tattooed on the collective psyche. In 2012, for example, a man named Richard Neill posted a message on the Facebook page of U.K. maxi-pad maker Bodyform, complaining that happy advertisements about managing menstruation misled him as he stepped into a sexual relationship with his girlfriend.

He wrote, and I am excerpting just the core of his message: "As a man, I must ask why you have lied to us for all these years. As a child I watched your advertisements [about] how at this wonderful time of the month the female gets to enjoy so many things. I felt a little jealous. I mean, bike riding, rollercoasters, dancing, parachuting, why couldn't I get to enjoy this time of the month. Dam [sic] my penis! Then I got a girlfriend...and couldn't wait for this joyous, adventurous time of the month to happen...you lied! My lady changed from the loving, gentle, normal skin-coloured lady to the little girl from *The Exorcist* with added venom and extra 360 degree head spin. Thanks for setting me up for a fall, Bodyform, you crafty bugger."

Bodyform responded by posting an online clip featuring a fictional CEO named Caroline Williams who replied directly, and playfully, to Richard Neill. In the video, the fictional CEO pours herself a glass of liquid tinted blue, looks straight into the camera, and says, "Hello, Richard...we read your Facebook post with interest, but also a sense of foreboding, and I think it's time to come clean. We lied to you, Richard, and I want to say 'sorry'....You're right. The flagrant use of visualization [in our advertisements] such as skydiving, rollerblading, and mountain biking — you forgot horse riding, Richard — are actually metaphors. They're not real. I'm sorry to be the one to tell you this, but there's no such thing as a happy period. The reality is, some people simply can't handle the truth."

At this point in the video, we see images of men in maxi-pad focus groups reacting in horror as they are presented with details that the CEO enumerates. She says: "The cramps, the mood swings, the insatiable hunger, and yes, Richard, the blood coursing from our uteri like a crimson landslide. So we knew we'd have to change our strategy...We have managed to maintain this illusion. But you, Richard, have torn down that veil and exposed this myth, thereby exposing every man to a reality we hoped they would never have to face. You did that, Richard. You. Well done. I just hope you can find it in your heart to forgive us." The actor sips from a glass of tinted blue water — the same sanitized stuff used countless times in advertisements about the absorbency of pads and tampons (blue, of course, appearing cleaner and less offensive than the colour of blood) — raises it as if to toast Richard, and lets out an unmistakable fart. She says, "Oh, sorry, Richard. You did know that we do that too. Didn't you?"

Culturally there has always been a taboo around bleeding women. An argument could be made that because women have been socialized to think of their own menstruations as unclean, it suits them to keep men in the dark about what they might be going through, or that they might be having a period, or what that might involve.

The *Globe and Mail* reporter Stephanie Nolen caught the essence of this strange dynamic in an article that she wrote last year about a man in India who discovered that his wife was forced to use unsanitary and make-shift means to catch the flow of blood during her period. Arunachalam Muruganantham found his new wife sneaking around the house one day carrying old rags and newspaper. He asked her what was going on, and she brushed him off. When he insisted, she admitted that she was menstruating. He asked her why she wasn't using sanitary napkins and she said that they couldn't afford them. Muruganantham embarked on a long quest to develop an affordable sanitary napkin for Indian women, a process which included parking himself outside a medical school and asking female students about their periods, and walking around with a blood-filled goatskin strapped to his body, which was connected by means of a tube to his underwear. From time to time, he would squeeze the goatskin and force it to release blood, to see if the napkin he had made — and was wearing — would absorb the flow.

Muruganantham's efforts to invent the perfect — and affordable — sanitary pad were judged sufficiently insane to prompt his wife, mother, and sisters to move out. They came back after he won a prize from the Indian Institute of Technology for his invention of a tabletop machine capable of shredding cellulose fibre and shaping it into sanitary pads with a hydraulic press. Instead of trying to make a bundle of money by selling his invention to a company, he sold it at barely above cost to some rural Indian women who started production in a rented garage. The women who purchased the equipment now make the pad and go door to door explaining its use and sanitary qualities, and Muruganantham also delivers the machines to isolated mountain villages so that girls in schools can make their own pads and make some money in the process. As Nolen said, miraculously, in a country where men are not known for meddling in the monthly affairs of their women, Muruganantham has become known as the sanitary-napkin man.

The American feminist and activist Gloria Steinem jokes about how different the world would look if it were men, instead of women, with menstrual cycles, in "If Men Could Menstruate," an essay in her book *Outrageous Acts and Everyday Rebellions*. To begin with, she says, "Men would brag about how long and how much." In addition, she predicts, "Generals, right-wing politicians, and religious fundamentalists would cite menstruation ('*men*-struation') as proof that only men could serve God and country in combat ('You have to give blood to take blood'), occupy high political office ('Can women be properly fierce without a monthly cycle governed by the planet Mars?'), be priests, ministers, God Himself ('He gave this blood for our sins'), or rabbis ('Without a monthly purge of impurities, women are unclean')."

These are but a few examples of how the bleeding that differentiates the genders gave rise to negative social stereotypes that still permeate our societal beliefs and values today. But sadly the divisive nature of blood does not end there.

The necessity of access to safe and affordable methods

of dealing with menstrual blood is not limited to developing nations. In 1980, epidemiologists began reporting cases of toxic shock syndrome related to the use of a super-absorbent Rely tampon that Procter & Gamble had manufactured for use in the United States. The tampon had been designed to contain a woman's entire menstrual flow without leakage or replacement. The Rely tampon was meant to contain nearly twenty times its weight in blood, and to expand into the shape of a cup as it filled. The company recalled the product, but later it was demonstrated that the super-absorbent tampons of any manufacturer were linked to increased risk of menstrual toxic shock syndrome. TSS, as it is known, results from bacterial infection. Although it can present in otherwise healthy individuals, it can lead to stupor, coma, organ failure, and death.

Men just don't have to think about these things. Many are proud to shed their blood in sport or war as a badge of courage and proof of devotion to a noble cause. But men do not have to consider how to deal with regular, healthy monthly bleedings. Jerry Seinfeld thought it was funny enough to crack this joke: "TV commercials now show you how detergents take out bloodstains, a pretty violent image there. I think if you've got a T-shirt with a bloodstain all over it, maybe laundry isn't your biggest problem."

The joke is custom-made for men, for whom blood in clothing often results from sport or war. But blood in clothing and laundry, for women, is a fact of life.

WE HAVE TRAVELLED A LONG JOURNEY in coming to understand the way that blood works in bodies. The things we can do with blood seem nothing short of miraculous. We know how to withdraw it, how to break it into its main parts (red blood cells, white blood cells, platelets, and plasma) in a centrifuge. We store it safely and transport it around the world. We treat and analyze it to ensure that it is free of diseases. We carry out millions of transfusions each year around the planet. We employ dialysis to clean the blood of a person whose kidneys don't work. We manufacture insulin and a variety of pills to help diabetics maintain acceptable levels of sugar in their blood. We have developed artificial blood-clotting products designed to prevent people from dying of bleeding disorders.

But there are still countless ways that our blood can betray us. As much as our blood works to regulate itself and help repair or maintain the health of our bodies, it can also circumvent its own rules, turning on us in potentially lethal ways. We have bleeding problems such as hemophilia and von Willebrand disease, cancers of the blood or bone marrow such as leukemia and lymphomas, and disorders such as anemia and sickle-cell disease, in which blood fails to carry oxygen properly from the lungs to the rest of the body. Blood disorders affect millions of North Americans each year, straddling all boundaries of age, race, sex, and socio-economic status. In addition, we are facing an epidemic of diabetes, a disease manifested by a surplus of sugar in the blood.

When sugar (or glucose) levels remain too high — either because the pancreas is not producing insulin or because the insulin is unable to do its proper job in controlling blood sugar levels — the body begins to break down. In the worst-case scenario, nerve endings fray, body extremities have to be amputated, organs begin to fail, and the patient dies.

An exogenous agent can also corrupt blood. Your body doesn't choose to break down, but a foreign visitor forces it to do so. This can take place through a variety of means, such as sex, blood transfusions, the use of infected needles, and mosquitoes.

In 1989, I was working as a volunteer with Crossroads International in the landlocked country of Mali, in West Africa, when I became aware that I was sick, and nauseated, and feverish, and that my bones were aching terribly. How to describe the symptoms: it felt like the worst flu I've ever had. I had been faithfully taking Aralen, an anti-malarial prophylactic — the most awful-tasting pill I have ever put in my mouth, by the way — but it did not prevent me from acquiring malaria. It is possible, however, that it kept the disease from becoming fatal. I took refuge in the house of my good friends Francine and Pierre Baril in Bamako. I let them take care of me, and I believe I took some quinine. Other than that, I lay under a ceiling fan and drank lots of water and waited out the flu-like symptoms. They held me in their grip for about a week, and then they let go, and I recovered.

Not every person is so lucky. The World Health Organization estimates that there were more than two hundred million cases of malaria in the world in 2010, and that more than six hundred thousand people died from it. Most deaths occur in Africa, where, according to the WHO, a child dies of the disease every minute. People commonly assume that malaria is restricted to tropical climates such as sub-Saharan Africa, but we find instances of it all over the world — in Asia and in the Americas too, for example, with cases showing up annually in the United States and travellers returning to Canada with the disease in their blood. Some face the misfortune of periodic new bouts of malaria over the course of a lifetime. Malaria is not as much geographically specific as it is climate-related, and if global warming continues, the disease could present itself more often in northern climates.

It took scientists a long time to figure out that mosquitoes are the vectors of malaria. In the eighteenth century, for example, malaria raged in the sea islands off the coast of South Carolina. It was so bad that the spring and summer were known as the "sick season" on slave plantations, and some Southern whites left their plantations entirely to be run by African slaves until the season passed. It was thought at the time that noxious airs were responsible for the fatal illness.

It was not until medical breakthroughs in 1898, when the Scottish physician Sir Ronald Ross proved the complete life cycle of malaria in mosquitoes — for which he was awarded the Nobel Prize for medicine — that we came to understand that the disease is passed from one human being to another by means of the mosquito. Malaria is caused by any of four different plasmodium parasites, transmitted by the female anopheles mosquito, of which there are about twenty key species around the world.

You don't have to touch the person who is infected. You don't have to meet the person, or even be in the same room. Presumably, the infected person could die after having been bitten by the mosquito, and it wouldn't matter to you. All that matters is that a female mosquito bearing sporozoites in its saliva glands chooses to bite you, and to spit into your bloodstream while it extracts a tiny hit of your blood. The disease usually shows up ten to fifteen days after the infective mosquito bite.

It is disturbing to stop and think about how malaria works, because the mosquito links the blood systems of people who don't even know each other. It doesn't stop with malaria. From human to human, mosquitoes can also transmit West Nile virus, dengue fever, yellow fever, and Japanese encephalitis. We are more connected than we think, and sometimes in dangerous ways. The more we learn about blood, the more we understand how all blood is hopelessly and forever intermingled, just like humanity itself, across culture, across gender, across age and race, and even across time.

SOMETIMES, THE VERY PERSON who is trying to keep you healthy can be the one who infects your blood and causes

your demise. In the mid-nineteenth century, while working as an obstetrician in the Vienna General Hospital, the
Hungarian physician Ignaz Semmelweis discovered that
new mothers died from puerperal fever (a form of blood
poisoning also known as childbed fever) at a far higher
rate in a maternity ward in which doctors worked, than
in a second maternity ward in which midwives worked.
The problem was publicly known to such a degree that
some pregnant women begged not to be admitted to the
ward supervised by doctors and preferred to give birth
in the streets. This was in the 1840s, about two decades
before the British surgeon Joseph Lister built on the
work of the French chemist Louis Pasteur and began
to require — with great success — doctors to wash their
hands and medical instruments between patients, to
reduce the spread of germs.

Working a step ahead of his time, Semmelweis postulated that doctors and medical students were picking
up "cadaverous particles" while dissecting cadavers, and
carrying them into the maternity wards. When he pre-
vailed upon physicians to wash their hands in a chlori-
nated lime solution between their autopsies and their
maternity ward work, the maternal mortality rate from
blood poisoning dropped dramatically. The resolution
of this problem became an obsession for Semmelweis.
He was convinced of the importance of his discovery but
failed to convince his peers of the necessity of assidu-
ous hand-washing to prevent rampant blood poisoning
in maternity wards.

Contemporaries derided the work of Semmelweis, who lost his job at the Vienna hospital. Doctors would continue to spread disease from cadavers to patients, or from patients to patients, causing new mothers and others to die of blood poisoning for many more years, until the germ theory was accepted and hand-washing became the norm. Semmelweis suffered from various professional setbacks and personal problems. He was tricked into visiting an asylum, which he was not permitted to leave. Instead, he was straitjacketed and beaten.

Within two weeks of being detained, it appears, Semmelweis cut his finger during an altercation with asylum employees. The wound became gangrenous. Semmelweis developed blood poisoning — the very problem he had been trying to avert in maternity wards. Within two weeks of being "admitted" into the institution, Semmelweis died. The year was 1865. It was not until the following century that his passionate efforts to help women avoid blood poisoning in maternity wards led Semmelweis to be dubbed the "saviour of mothers."

Likewise, Bruce Chown of Winnipeg could be considered the modern-day "saviour of babies." Over the course of time, countless fetuses and newborns have died because — as we finally came to understand near the middle of the past century — their blood types were incompatible with that of their mothers. This is because an antigen, or molecule on the surface of the red blood cell, is present in most, but not all people. A person with this antigen is considered Rh-positive and will have a positive blood type. Someone without it is Rh-negative and will have a negative blood type. With regard to pregnancy and childbirth, this factor becomes critically important — and potentially fatal for the fetus or infant — when the mother is Rh-negative and the baby is Rh-positive. This can occur if the father is Rh-positive. If some of the baby's red blood cells leak into the mother's bloodstream — and this can happen during childbirth or in the case of miscarriage, abortion, or other in-uterine trauma — the mother will develop antibodies against the Rhesus factor in the baby's blood cells. As a result, if the mother becomes pregnant again with an Rh-positive baby, the mother's antibodies may attack the baby's blood. This can result in jaundice, brain damage, and the death of the fetus or newborn child. Indeed, as recently as the early twentieth century, mortality from erythroblastosis fetalis — also known as hemolytic or Rh disease — was 50 percent. It accounted for 10 percent of fetal and neonatal deaths in Canada, and claimed the lives of some 10,000 babies annually in the United States.

There are now almost no such deaths in Canada, partly as a result of the pioneering work of Bruce Chown. In 1944, Chown — head of pediatrics at Winnipeg Children's Hospital — founded a lab in the basement of the hospital to study the cause of Rh disease. Chown, who is described by the Canadian Medical Hall of Fame as one of the greatest Canadian physicians of all time, developed the first Rh exchange transfusion in Winnipeg in 1945, and went on to develop a facility with

Connaught Laboratories to manufacture an Rh immune serum. By coating the fetus's red blood cells and thus disguising their foreign nature, the serum tricks the mother's bloodstream into thinking it's compatible with that of the baby. Rh immunoglobulin was licensed for use in 1968, and soon helped prevent many more fetal neonatal fatalities in Canada and around the world. The Rh immune serum is developed from the donated plasma of an Rh-negative donor whose blood has developed Rh antibodies as a result of being exposed to Rh-positive blood cells.

The discovery of the Rh immune serum spurred again the drive to donate, and provided an outlet for the human need to save another's life. Raymonde Marius, who was seventy-nine when I spoke with her in April 2013, gave plasma more than one thousand times over the course of forty years. After giving birth to five healthy children, she experienced three miscarriages. In the case of the final two losses, which took place in the 1960s, Marius had to carry her fetuses to full term and deliver babies that she already knew had died. The deaths occurred because of the Rhesus disease: her blood type was incompatible with that of the fetuses growing in her womb.

After her own personal losses, Marius wanted to prevent other mothers from having miscarriages as she did. Because she had lost babies to Rh disease, her blood was rich with antibodies that could be used to create an immune system to give to other Rh-negative mothers who were likely to have Rh-positive babies.

When a doctor asked if she would consider donating plasma to be used to manufacture an immune serum, she entered a donation program that lasted four decades. She spent one evening a week, every week, donating plasma. "I started donating at the age of about thirty-seven and donated for forty-two years," Marius told me. "It takes time to donate plasma. They take it. They spin it [in a centrifuge]. They remove the plasma. They take what they want and give us back the rest. And they do that twice. I was a fast blood donor. I had good veins. It took me sixty to seventy-five minutes to donate plasma.

"I got a plaque after the thousandth time (I donated blood). It was a nice plaque with my name on it. And I kept on donating after that." Marius said that she was initially paid $5 per plasma donation, and that the amount later rose to about $60. She said they were paying for her time and for her travel expenses, not for her blood.

Cheryl Lawson is an employee at Cangene Corporation in Winnipeg, which pays donors to give plasma that is used to manufacture WinRho SDF, an immune serum globulin used to prevent fatalities due to Rh disease. Lawson estimates that Marius may have donated as many as 1,500 times before she finally stopped. The plasma donation program began in Winnipeg in the 1960s and had an active corps of twenty or thirty female donors at the time. Now the company has 102 regular donors — some male, but mostly female — who commit to a regular, long-term donation program, according to Lawson. The blood of the donor frequently has to be readied for donation. A donor may have to receive several blood transfusions to stimulate their own blood to produce the antibodies necessary to develop the serum. Clearly, all this work — undergoing blood transfusions to become a successful donor, and then donating regularly for years — stems from the desire to give. When I asked why she donated plasma weekly for forty years, Marius had a simple answer: "I was helping another mother. Without my plasma, she wouldn't have been able to have a healthy baby."

Blood keeps us alive, and — right down to our red and white blood cells — has the ability to save us. But as we have seen, it is dangerous too. It can turn against us, and it can turn against those we love most deeply too. The Bava Batra, the last of the three tractates of the Talmud, might indeed be prescient and wise beyond all reckoning when it warns, rather ominously: "I, blood, am the cause of all illness."

AS THE AUTHOR OF *Something Happened*, a novel in which the main character worries, as only a hypochondriac will do, about how his body might do him in, the late Joseph Heller might have identified with this observation: there are so many ways that our blood betrays us! Some cancers involve a proliferation of white blood cells, and are generally treated by trying to destroy those same cells. Bleeding disorders are treated by blood transfusions, and by offering patients products to help their blood clot. Although not life-threatening by any means, and certainly not of the same order of magnitude as cancer or hemophilia, erectile dysfunction can certainly affect the quality of a man's life, and can be treated by a number of means — pills, injection, implants — all of which aim to move blood into the penis. We can even inherit defects of the blood from our ancestors, or betray our own descendants by passing them along.

My son, Andrew, eighteen years old and winding up his first year at university as I write, is an active, slim man. He played hockey and soccer throughout his childhood. He goes to the gym. Gets on a bike. Likes to go running. When we were younger, he and my two eldest daughters, Geneviève and Caroline, would sometimes run with me. Some of my fondest memories involve running five-kilometre community races with them, on Father's Day and on Thanksgiving. When they were young, I would run with whichever one was slowest in that particular race, so he or she wouldn't feel alone. Now I wouldn't be able to keep up with Andrew.

I had an ulterior motive for running with the children, Andrew especially. Sure, it's good for the body, and relaxing for the soul. A great stress reliever, and something I have loved to do since my own childhood. But I took Andrew running for the same reason that I have tried to warn him, ever so gently, not to overdo the desserts. Perhaps not to stack an ice cream scoop on top of a piece of apple pie, all of that having followed a piece of cake. Teenagers eat like that, and many can get away with it. But I have worried about Andrew. He is in a long line of diabetics, and it shoots like an arrow through the male line of my family.

My grandfather, father, and older brother all developed diabetes at forty-three. When I passed that age, I counted my blessings. Maybe the disease had skipped over me. Perhaps all those years of running had saved me. But I was to have no such luck. I was diagnosed two years later. My grandfather and father died of complications related to diabetes.

Diabetes stands out as one of the most common defects of the pancreas, and thus the blood, to be passed from one generation to another. It has reached epidemic proportions in Canada and other countries. Diabetes associations in North America estimate that some twenty-six million Americans and three million Canadians suffer from the disease. It manifests itself in three basic ways. As the Canadian Diabetes Association notes, Type 1 diabetes usually shows up early in life and results from an inability of the pancreas to produce the hormone insulin, which regulates the amount of glucose (or sugar) in the blood. Type 2 diabetes presents itself more often in adulthood (although more children are being affected as the epidemic grows), and involves an inability of the pancreas to produce sufficient insulin or an inability of the body to use the insulin well. Finally, gestational diabetes occurs during pregnancy and is considered temporary.

The consequences of diabetes can be lethal. Excessively high levels of blood glucose eat away at our nerve endings, attacking the organs and the body's extremities. This can lead to blindness and foot and leg amputations. It can also lead to kidney failure and death. Some people call the disease "the silent killer," because diabetics do not generally experience significant pain or discomfort in the early phases of the disease. It's a misnomer, however, because if it proceeds untreated or is poorly managed, the body will break down in the most excruciating ways, requiring operations, amputations, and dialysis.

I watched my father, Daniel G. Hill III, fall apart limb by limb. He had a wonderful, active, public life through most of his career as a sociologist, human rights activist, and writer about black history in Canada. His last years on the job were tough, but he made it through his final working stint as ombudsman for the province of Ontario. It was a five-year posting. I attended his swearing-in party, in 1984, and I remember driving him to the Ontario Legislature for it. He was still well enough to travel with my mother, go to work, and do battle on behalf of those who felt they had been treated badly by the Ontario government.

But he had good days and bad days, back then, and the day of his swearing-in was one of the bad ones. He was surely excited, and proud, and perhaps a tad anxious, going to the party. He called it glad-handing, when he went to social events and had to chat with a hundred well-wishers. He was the first black man to become the ombudsman of Ontario, and years earlier he had become the first director of the Ontario Human Rights Commission, and he was surely conscious of the social weight on his shoulders. He knew he had better not screw up on the job, do something stupid, or cause some scandal, because if he did the whole world would be looking to point fingers. To the best of my knowledge, he never screwed up on any job — certainly never enough to draw public opprobrium, or to jeopardize his own position.

I could see that he was worked up, in the passenger seat of the sedan that I drove to get him to the Legislature, but I was not prepared for what I saw. He became confused, irrational, demanding things that made no sense, and he appeared to be physically trembling. It upset me to see my father so out of it, on a day that he was expected to be in command of his faculties and to excite his well-wishers about the job that was ahead of him. He had invited many of our neighbours to the reception, and I recall that he was especially tickled and proud about that. With all the responsibilities that were about to fall on his shoulders as the person appointed to probe into accusations of wrongdoing by the very government that employed him, the thing my father obsessed about in the car was the neighbours: who was coming, if they had received their invitations, how they would react. This fussing on his part told me that something was wrong.

He got out of the car and started wandering off toward an employee of the Legislature who stood at the door, barking some sort of insane order. We got him inside and into a quiet room, and figured out that his blood sugars were probably low, and gave him a glass of apple juice. Fifteen minutes later, he was back to normal.

The irritability, irrationality, the trembling — all of this resulted from the amount of sugar in his blood. In the language of diabetics and their families, he was "having a low." Glucometers — pocket-sized devices that measure blood sugar — were bigger, more cumbersome, and slower in the 1980s than they are now. You had to wait a while for the number to show up on a little screen, although it now takes all of five seconds to get the reading. I can't remember for sure what his reading was that time, but it gave us all the information we needed to get the juice into him.

The amount of sugar, more properly called glucose, in his blood was low because he was injecting insulin daily, and had evidently miscalculated. The insulin drove the sugars out of his blood and into his cells, and did an overly enthusiastic job of it, and so he found himself trembling and disoriented. I know the feeling. Since I developed diabetes, I have had this experience many times. Let me tell you what it feels like: the life is draining out of you. Your bodily energy is like a sink filled with water, but somebody has pulled the plug and your very vitality is slipping away. You may feel melancholy. You may have the shakes. If you're sufficiently low, the shakes can be so bad that you have trouble pricking your fingertip to squeeze out a drop of blood so the glucometer can tell you just how low you are. Sometimes you feel so wretched that it does not even occur to you that you are low. But eventually you figure it out. I know all the numbers. Let me tell them to you. If you are eight or above, you're definitely too high. Not good. High blood sugar levels can lead to nerve and organ damage. If you are between six and eight, it's not so bad, but not optimal. When you are fasting, if your blood glucose reading lands between four and six, you're spot on. Where you are supposed to be. Where a healthy person with a healthy, functioning pancreas would be. Although for me, I start feeling bad at anything under five, and distinctly awful if I am at a four. If I am at three, I would probably be shaking so badly that I could not prick my own finger. Under two, and I would not be conscious. I've had one or two readings over the years between 2.9 and 3.1, but thankfully they've become rare as I've become more adept at anticipating what food, exercise, and insulin mix I need to stay in a better zone.

I usually guess my blood sugars before I take them, because I'm that kind of guy. Make a game of it, run the numbers in my head, try to guess — by what I've eaten, how much I have exercised, how many metformin pills I have swallowed, or how much insulin I have injected — just exactly what my levels will be. What I most like doing is guessing how much my blood sugars have dropped over the course of exercise: say, after going out for a one-hour jog. In my case, they can easily go down five points. I could start a run with a blood glucose reading of ten, and finish it at five. That would be quite typical for me. Sometimes I drop more, or less. Sometimes I finish a run higher than it was when I started. That's either because the liver kicked in and drove some glycogen into my bloodstream, or because I drank too much Gatorade in the course of the run. That's the other thing. I don't like to be anywhere without access to two things: my glucometer and some apple juice, in case I am having a low. If my blood sugar is going to betray me, I want to be prepared.

I don't mind watching my blood rise and fall over the course of a day. Doctors and nurses tell me that keeping close track of my blood glucose levels is one of the best ways to keep it in a healthy range. You'll know very fast if you've been eating too much, or injecting too much insulin, or failing to hit the right balance of food, exercise, and drugs. Other things, of course, can throw you off track too. Stress drives up the blood sugar count. So does being sick, such as having the flu. Although there is no cure for diabetes, millions more people would die or be incapacitated had it not been for the work of the Canadians Frederick Banting (who won the Nobel Prize for his life-saving research), Charles Best, and other scientists, who discovered that the use of insulin produced by the pancreas could stave off the ravages of the disease by helping diabetics keep their blood sugars at healthier levels.

Every part of our body craves blood. No part of us can do without it. But the body is a sensitive beast. It is humbling and striking to realize how acutely the body reacts to the tiniest fluctuations in the chemical composition of the blood. Through the natural physical processes of blood, we see the march of human life. We see fathers passing blood legacy on to sons; mothers mourning their unborn children when the blood mistakes the fetus for a foreign body; a nation in a test tube of blood — equipped with citizens working to keep the "economy" circulating, a government to maintain our infrastructure, an army to defend against foreign attackers. Blood, indeed, is the stuff of life.

BLOOD IS MEANT TO REFLECT our humanity. In the Old Testament, the lamb's blood on the doorposts of the Hebrews signified the innocence and the vulnerability of the Israelites in the time of Egyptian slavery. Indeed it inspires the Jewish tradition of Passover, celebrated by millions today. A soldier, having been injured on the battlefield but having survived to return home later, is said to have shed blood for his or her country — and this is meant to confer valour and honour on the soldier. We are supposed to respect him or her more, precisely because they have lost blood for their nation, or for the goals of their people.

We now know that people have different blood types, but the fundamental sameness of our blood — its colour, texture, and functions — is supposed to link us as human beings. This is what William Shakespeare says in creating the character Shylock in his play *The Merchant of Venice*. Shylock, a Jewish moneylender, is intent on

extracting a pound of flesh from the defaulting debtor
Antonio, who has previously insulted him. However,
as the play unfolds, the tables turn against Shylock. He
learns that he will not be allowed to cause any loss of
blood in taking Antonio's pound of flesh. In addition,
Shylock is accused of conspiring against the life of a
Venetian citizen. When his situation looks desperate
and he is anxious to establish that he is just the same as
those who would accuse him, Shylock — a Jew — says to
Salarino: "Hath not a Jew eyes? Hath not a Jew hands,
organs, dimensions, senses, affections, passions? Fed
with the same food, hurt with the same weapons, subject to the same diseases, healed by the same means,
warmed and cooled by the same winter and summer, as
a Christian is? If you prick us, do we not bleed?" Shylock's
plea does not save him from defeat and humiliation: he
must forfeit his property, will his estate to a daughter who eloped with a Christian suitor, and convert to
Christianity.

Blood filters through our consciousness thanks to
Shakespeare and countless other writers, and we feel its
rhythms the same way we sense the natural beauty of
the iambic pentameter. Blood resides in the drumbeat of
our hearts, but it is also deeply embedded in the tempo
of our language.

What, after all, is a bleeding heart but a contemp-
tuous reference to a person who cares too quickly and
superficially for the plight of others? If you are bloody-
minded, you concern yourself overly with gruesome

matters. When we think of blood sports, we have in mind the activities that celebrate not just killing but the spilling of blood: the way foxes might be torn about in a hunt, or the way picadors stab a bull so many times that it is weakened and all its blood streams across its ribs and onto the earthen ring in the bullfight. A crowd that calls out for blood is your collection of rabid hockey fans when a scrap breaks out on the ice, or the instantly formed circle of high school kids around a pair of fighting students, each group hungry to see blood spill. If you are a cold-blooded murderer, you are no longer quite human but rather have been likened to a reptile with an entirely different — and distinctly unhuman — blood system. If you have blood on your hands, well, you've done something terribly wrong and people have either figured it out or they surely will soon. Blood on the hands pretty well guarantees that people will want to see you brought to justice. We all know that getting money back from the tax collector who was, perhaps, too exuberant in the exercise of his duties is like extracting blood from a stone. If you have a blood brother, you have opened up your flesh to that of a friend who has done the same, and in exchanging bloody skin surfaces you have created a bond that runs as deep as — or perhaps even deeper than — family. If you have bad blood in your family, everybody knows that it would be foolish indeed to marry you, because you obviously fall short of having a respectable moral code. If you are out for blood, you have violent intentions and you should know better.

Let's hope that someone brings you to your senses, and that blood stops pounding in your temples. If you have royal or so-called blue blood, millions of people will look up to you and possibly agree to keep paying taxes to keep you in Buckingham Palace, and in our collective imagination, your corpuscles are indeed of an elevated, unusual, and vaunted quality. If somebody makes your blood boil, you have become so upset that you have temporarily lost your humanity and become another sort of animal entirely, whose blood is allowed to change temperatures radically. If somebody says of you that "blood will out," you know it is an insult, to the effect that your family characteristics cannot be concealed. In a sonnet called "Blood," first published in the *Nation* in 1928, the American poet Robert Frost wrote: "Oh, blood will out. It cannot be contained."

Blood, indeed, filters into every aspect of our language and defines who we are: in our emotional states, in our social ranking, in our state of innocence or moral guilt, and most important of all, in our relationships to each other.

Blood is truly the stuff of life: a bold and enduring determinant of identity, race, gender, culture, citizenship, belonging, privilege, deprivation, athletic superiority, and nationhood. It is so vital to our sense of ourselves, our abilities, and our possibilities for survival that we have invested money, time, and energy in learning how to manipulate its very composition.

TWO

WE WANT IT SAFE AND WE WANT IT
CLEAN: BLOOD, TRUTH, AND HONOUR

I FELT A WAVE OF EMPATHY when I watched Paula Findlay finish last in her triathlon at the 2012 London Olympics. The Canadian crossed the line crying and in obvious discomfort. Findlay is an elite athlete. Prior to the Olympics, she had won numerous international triathlons, been ranked number one in the world, and considered a medal favourite in London. General Mills, the cereal company, was using her image on boxes of Reese's Puffs. However, on an August day in London's Hyde Park, Findlay, twenty-three at the time, was the last of fifty-five female competitors to cross the line after a 1.5-kilometre swim, a 43-kilometre bike ride, and a 10-kilometre run. Posting a post-race description on her blog, Findlay mentioned that her swim went poorly and her bike ride even worse, but that she felt wobbly and powerless after running the first of four 2.5-kilometre laps. "I pulled off to our team doctor, crying that there was no way I could physically finish three more. He encouraged me to pull myself together and finish if I could, I'd be more satisfied with crossing the line than not. So I ran three of the most painful, embarrassing laps ever, being lapped by the race that I was supposed to be a contender in, humiliated and screaming at myself inside."

Findlay had recently come off a hip injury, but another problem contributed to her poor showing at the Olympics. In a blog post five weeks after the race, Findlay wrote: "I had some blood work done about a week after I arrived just to make sure that everything was normal. I was feeling tired but assumed this was just an effect from training hard again. Unfortunately the numbers came back with some of the lowest iron levels that the doctors had ever seen. It is a simple but quite serious problem that likely had a huge impact on my race in London, and got overlooked because of the focus on healing my injury."

At the time that this book was going to press, Findlay appeared to be not just healthy and fit but a threat to her competitors again. In March 2013, she won one triathlon in Florida and another in Austria. Competitive sport is unforgiving. Bringing a serious iron deficiency to the starting line of the world's most competitive triathlon is akin to missing an organ or a limb. You need the iron to produce the blood cells to carry the oxygen to keep you in the race. Minus iron, you will finish in the position that someone, from some country, must occupy: dead last. Findlay competed with honour at the 2012 London

Olympics. But she would have had more success had her blood been in order.

Watching the Olympics, I wanted to reach through the TV screen and give Paula Findlay a hug. I thought her courageous to complete the race. I believe that it takes as much courage to suffer and finish last as it does to vanquish all your competitors and run away with the gold medal. I am not an elite athlete and have never been one, but I could certainly identify with finishing last in a running race — something that I had experienced many times as a teenage middle-distance runner.

I began to dream, at the age of eleven, of becoming a champion runner. By the time I joined a track club at the age of twelve, I had a plan in place. The year was 1969. One year earlier, American 200-metre sprinters Tommie Smith and John Carlos had raised their fists to make the black power salute as they stood on the podium with their gold and bronze medals at the Mexico City Olympics. They were expelled from the Games and vilified back in the United States, but I loved them for their daring chutzpah and for their fearlessness about expressing black pride. I had no thoughts about imitating their protests, but certainly wanted to achieve the same degree of success and fame. The blueprint was clear. By the 1984 Olympics, by which time I would be twenty-seven, I expected to win the gold medal in the Olympic 5,000-metre race. I would hang behind the race leaders for the first 3,000 metres, surging to break their will (and empty their lungs) until the 4,500-metre mark. At which point

I would accelerate again, steadily pulling away from my last rival over the final 150 metres. I would cross the finish line 20 metres ahead of the next runner.

This grandiose dream was all to please my father, although I didn't understand it at the time. A hard-working, domineering, charismatic, brook-no-dissent-at-home African-American immigrant to Canada, my father had little interest in relaxation, except when it came to turning on the "boob tube," as he called it, to watch westerns and sports. While he whooped and hollered at the athletes on the TV screen, I studied him. Dad had three sporting passions, all left over from the near-religious worship of sport that had marked his own upbringing in the United States: boxing, football, and track and field. In our household, two athletes in particular became my father's obsessions: the heavyweight boxing champion Joe Louis, also known as the Brown Bomber, and the track and field star Jesse Owens. Both sprang into international prominence as a rebuttal to the notion of Aryan superiority in Nazi Germany, Louis by trouncing the boxer Max Schmeling in 1938, and Owens by winning four track and field gold medals in the 1936 Summer Olympics in Berlin.

I decided that the most effective way to secure my father's everlasting admiration was to become the most successful runner in Canadian history. By the age of thirteen, I was training daily and keeping a detailed log-book of my workouts. By fourteen, I often trained twice daily, getting up to run five or ten kilometres before school and doing intervals on the track in the afternoon or evening. I undertook my first very long runs by the age of fifteen, twice running the entire thirty-two-mile Miles for Millions — a fundraiser for charities that was massively popular in Toronto in the 1970s. Thousands of Torontonians hit the streets once a year to walk the thirty-two-mile route. I chose to run it. The second time I ran it, I believe I was one of the first participants to finish. But I am not entirely sure how many people finished ahead of me. It was meant to be a walkathon, not a race, and the walkers were spread out along the streets of Toronto, many hours behind me. I arrived at City Hall so early that nobody was there to greet me. Thousands of people who had the good sense to walk the route crossed the finish line over the next twelve hours — long after I had taken the bus home, had a bath, bandaged my blisters, eaten a bowl of ice cream, and gone to bed.

After Miles for Millions, I continued to dream of Olympic glory. Kipchoge Keino, the Kenyan superstar whose strength was said to derive partly from drinking cow's blood, had won the gold medal in the 1,500-metre race and the 3,000-metre steeplechase at the 1972 Olympics in Munich. I thought about Keino and renewed my private vow to achieve my goal. I trained through sore shins and blood-filled toenails, and although I was getting faster as I aged, my competitors were progressing much more quickly. At fifteen, I was no closer to winning any significant races than I had been at the age of twelve. When I went for Sunday training runs on hilly country roads in southern Ontario with the other teen-agers in my track club, I would soon be left far behind the others to ward off unleashed country dogs alone. What must have been patently obvious to my coach, friends, and parents, but which took time to dawn on me, was that I did not own the body of a person destined to be, or capable of becoming, a great or even a good athlete.

Finally, when I was sixteen or so, my track coach — David Steen, a *Toronto Star* reporter who had twice won the gold medal in the shot put at the Commonwealth Games — suggested that I take an "oxygen uptake" test to determine how effectively my blood transferred oxy-gen to my muscles. It was an unpleasant stress test, car-ried out on a stationary bicycle at the Fitness Centre in Toronto. While pedalling to exhaustion, I had electrodes taped to my chest and a mask locked over my mouth. By measuring the air that I expelled, the test would reveal how well I processed oxygen. I was a skinny, ultra-fit teenager who was logging about seventy kilometres a week in training runs, but the test did not reveal the ath-letic profile of a runner.

In writing this book, I asked Steen about his memory of this test. He said he thought it indicated that I had 75 percent of what would be considered normal cardio-respiratory capacity. I remember words a tad more blunt. I recall him saying, in a playful but concerned way, that the test suggested that I had the lung capacity of a forty-year-old smoker. Our memories coincide on one detail: the results were so pathetic that Steen, whom I loved and

admired, encouraged me to give up running and special-
ize in English literature.

The suggestion wounded me, although I now know
that it blended love with insight. At the time, I understood that I would have to change the way I approached
running. I would still train and compete. But I would no
longer dream. I discovered that it was an absolute, utter
lie to say that any person can accomplish any goal if only
they set their mind to it. I had run into the limits of my
own blood. It could carry and deliver oxygen only so
well. Well enough for me to develop decent calf muscles
and to finish Miles for Millions hours ahead of legions of
walkers. But not well enough to beat any serious runner
on the track. I learned to focus, instead, on the pleasure
and calm that came from pushing myself physically. I
would make my heart pump for its own sake. For the
joy of exercise, and for the sense of accomplishment.
I did not have the body of a champion runner, and no
amount of training would enhance my lungs, heart, and
blood enough to get me there. I had to honour myself by
understanding my own blood better and adjusting to my
limitations.

In this chapter, I will explore two means by which we
associate blood with notions of truth and honour. The
first concerns what we do with our blood. How do we go
about removing it from human bodies, and once it is out,
what do we use it for? This initial section of the chapter will explore how we offer blood to the gods, slay our
oppressors, depict heroic violence in art, and sometimes

fail as artists by neglecting to pay enough respect to blood. This first part will also examine three issues that are both social and medical: stem cell research, the tainted blood scandals erupting in many countries in the 1980s, and blood donation policies. Each of these issues pertains to notions of honour and truth with regard to our blood. Because they deal with the very core of our bodies, the debates on these subjects have been vociferous.

In the second part of this chapter, I will focus on physical changes that are invisible to outside eyes, and even to our own: adjustments to our blood composition. Such adjustments mark us as profoundly as plastic surgery. Changes to the blood in our veins have the power to affect our self-perception and how others see and treat us. Do you want to know about a person's honour or truthfulness? Look closely at their blood, and how its composition may have been altered.

HOW DO WE CONVINCE the ones we love, and the ones we wish to impress, that we deserve their affection and support?

We work extra hours for our bosses when we truly wish to impress them, and we bust our backsides to render good service or turn in a great product. In exchange, we hope we will be protected, remunerated, and promoted.

To earn the trust and love of children, we hold them and show up when they need us. In the kitchen, on the street, in the doctor's office, at the daycare or school drop-off or pickup, we secure our children's trust by offering daily acts of support, and by respecting their individuality.

How do we keep the ones we love, romantically? We show an interest, court like crazy, and when they finally say, "Yes, I will love you," we find ways to love them back daily, monthly, and yearly. For some of us, this means providing money and security. For others, it may suggest meals and other support offered at home. For relationships that satisfy, we hold each other whenever we can, and most certainly through the darkness of night. We look out for each other. We make love to move into as intimate a space as possible with our loved one. Emotionally and physically, we touch each other to preserve and nurture our love.

So how does that work for our relationship, if we have one, with our God or gods? Over the millennia, we have principally looked to one means of connection with our deities: we spill blood. Blood of animals, of other humans, or our own.

Slicing open a vein, bleeding an animal properly, or even ripping out someone's heart and holding it up to the sun is the ultimate way to tell our God: we fear you, and we hereby prove our fealty and ask that you meet our needs today and tomorrow.

Would we offer God a bag of peanuts, or $10 from a $200,000 bank account? The idea is risible. To win the care of God, we are meant to lose what counts the most: human blood, and sometimes human life.

From Biblical writing and earlier, we have long associated bloodshed with honour, expiation, and offering.
Human sacrifice — with blood as a central element —
seems to date back as far as human civilization. The
Mayans, who settled some four thousand years ago
in what is now southern Mexico and northern Central
America, practised some forms of human sacrifice. In
the ancient culture of the Zapotec peoples, who settled
around 1150 BCE in what is now the Oaxaca area of
Mexico, two types of blood sacrifices were practised:
priests cutting themselves during cult ceremonies, and
ripping the hearts out of prisoners or slaves to offer life
force to the spirit powers. Historians and others have
noted that human sacrifice has been carried out in many
corners of the world: in ancient Rome, ancient Sumeria,
China, and the West African kingdom of Dahomey, to
name just a few.

The Aztecs, who lived in Meso-America from about
1400 to the time of the arrival of the Spanish explorers, are perhaps best known for practising human sacrifice. As Mark Pizzato describes in *Theatres of Human
Sacrifice: From Ancient Ritual to Screen Violence*, Aztec
sacrifice rested on creation myths having to do with
gods who sacrificed themselves to allow a new sun to
rise over the empire. Because the gods had died for the
sun to ascend, humans too would have to be sacrificed
to ensure that the sun would continue to rise each day
and feed mankind. The Aztecs, Pizzato writes, sacrificed
children or prisoners of war on a nearly monthly basis.

Victims were schooled with great care and detail. In their daily interactions with other people, they assumed the characteristics of a god so that the final moment of their lives — when they were held down so that a priest could extract their beating heart and offer it up to the sun — would approximate the sacrifice of the early gods.

The notion of offering something of primordial value to the gods, in exchange for protection, or food, or the continuation of the rising sun, is also rooted in religious texts. Rembrandt and Caravaggio, among other painters, both explore on the canvas the most famous near-sacrifice in Biblical literature: the story of Abraham and his son Isaac.

According to the Book of Genesis, God commanded Abraham to take his son Isaac up Mount Moriah and to sacrifice him. Bob Dylan comments on this act in his song "Highway 61 Revisited," suggesting with his colourful lyrics that Abraham might have asked if this were some sort of joke. However, it was not uncommon in the time of Abraham for children to be sacrificed to pagan gods. According to Genesis, Abraham raises the knife to obey God's command. An angel appears and says: "Lay not thine hand upon the lad," and tells Abraham that it was merely a test of his faith.

As Steven Shankman observes in his book *Other Others: Levinas, Literature, Transcultural Studies*, the works by Caravaggio and Rembrandt offer contrasting interpretations of this key moment. In his 1603 painting *The Sacrifice of Isaac*, Caravaggio, who came first, showed Abraham with the knife in hand, holding the groaning face of his young son, at the moment when the angel grabs hold of the father's wrist and prevents the sacrifice. Rembrandt rendered homage to the same Biblical story in his 1635 painting *The Sacrifice of Isaac* and in his 1655 etching *Sacrifice of Abraham*. Here, I'll refer to the former work. Rembrandt painted it at the age of twenty-nine, the same year that his first son died in his infancy. In the painting, Abraham uses his giant left hand to smother the face of his son at the moment when the intervening angel knocks the knife completely out of his hand. Isaac's bound body is exposed in Rembrandt's painting. Naked but for a cloth around his waist, Isaac is on his back, knees bent, hands pinned beneath him, completely vulnerable and immobile as heavenly light spills across his body. In that bright, surreal light, blood gushing over Isaac's skin would render even more horrid the barbarity of a father killing his own son.

In Genesis, Isaac is saved. When the angel intervenes, Abraham finds a ram caught in a thicket and slaughters it instead. According to Genesis 22:1, God speaks to Abraham through the angel and says, "Because you have done this, and have not withheld your son, your only son, I will indeed bless you, and I will make your offspring as numerous as the stars of heaven and as the sand that is on the seashore. And your offspring shall possess the gate of their enemies, and by your offspring shall all the nations of the earth gain blessing for themselves, because you have obeyed my voice." The story is meant to signal the end of human sacrifice as condoned by God. From this moment forward, as far as blood sacrifice is concerned, animals will suffice.

To this day, the ritualized spilling of blood permeates humanity and evokes notions of honour. Japan has a long tradition of self-sacrifice for the sake of honour and country. The ancient warrior class known as the samurai practised seppuku, a form of ceremonial disembowelment, as a way to die with honour or to atone for having shamed themselves or their nation. Some women — wives of the traditional samurai — have also committed ritualistic suicide, by slicing the arteries in their necks. At times, this was done to avoid capture or rape at the hands of invading enemies.

In English, the word *kamikaze* refers to recklessly self-destructive behaviour. It entered the North American imagination as a reference to the World War II Japanese fighter pilots who committed suicide by crashing their planes into enemy ships. Thousands of Japanese pilots committed suicide in this way, and succeeded in sinking dozens of Allied vessels. The attacks have often been interpreted as a manifestation of the emphasis on loyalty and honour in Japanese culture. Writing for the BBC History website, journalist David Powers quotes a twenty-three-year-old Japanese airman writing home to his mother just days before his suicide mission in August 1945. Ichizo Hayashi wrote: "I am pleased to have the honour of having been chosen as a member of a Special Attack Force that is on its way into battle, but I cannot help crying when I think of you, Mum. When I reflect on the hopes you had for my future...I feel so sad that I am going to die without doing anything to bring you joy." Honour may have been central to Japanese culture, including honour in death. But that does not mean that every soldier leapt with purpose and joy toward that honourable death. Some, it is safe to bet, were surely shoved. Forced to be honourable, and to die for their country.

Committing suicide to preserve one's honour is one thing, and murdering someone to accomplish the same goal is quite another. The United Nations estimates that some five thousand honour killings — the murder of relatives, often women who are said or believed to have dishonoured their families by having unauthorized sexual or romantic relationships, or who refused arranged marriages — take place each year. It is a method of exacting conformity to rigid rules that men set for women, and punishing those who won't cooperate. In the eyes of those who practise honour killings, it removes the "stain" from a family's blood and intimidates others into accepting coerced behaviour. In other words, "Do as I say or I'll kill you."

Writing for the *Independent* in 2010, Robert Fisk noted that many women's organizations believe the number of honour killings to be at least four times higher. It's a barbaric act and a perversion of the term *honour*. As Fisk wrote, "The details of the murders — of the women beheaded, burned to death, stoned to death, stabbed, electrocuted, strangled and buried alive for the 'honour' of their families — are as barbaric as they are shameful." Fisk focused on honour killings in the Middle East, but acknowledged that the contagion has spread to Western nations. Indeed.

Writing in 2011 for the *Canadian Criminal Law Review*, University of Sherbrooke law professor Marie-Pierre Robert noted that the number of honour killings reported in Canada is on the rise. The first was in 1954, the second in 1972, and the third in 1983, Robert wrote, but then another thirteen took place between 1999 and the date of her article, in 2011. In 2012, a jury convicted a Montreal couple and their son of murdering the husband's three daughters and his first wife because the daughters had been dating, refusing to wear traditional Islamic clothing, and skipping school.

It would be smug and self-serving to attribute honour killings solely to Islamic culture and to leave the conversation there. In nations founded on Judeo-Christian values, murdering one's spouse or partner can be considered a Western equivalent of honour killing. Often, it involves homicide carried out by a man who wishes to control a woman and does not approve of what she says or does. According to Statistics Canada, between the years 2000 and 2009, 738 spousal homicides took place in Canada. Women were three times more likely than men to die of spousal homicide. According to a scholarly article written by Bridie James and Martin Daly in a 2012 issue of the journal *Homicide Studies*, police reported 13,083 cases of spousal homicide in the United States between the years 1990 and 2005. Of the spousal homicides, 9,828 of the victims were women and 3,255 men. Whether we think of honour killing in some cultures or family violence in others, spilling a woman's blood to control her behaviour is widespread around the world.

BLOOD AND HONOUR FREQUENTLY intermingle, even when death is not involved. The gods of Greek mythology were said to have a substance known as ichor in their veins. This gold-coloured fluid was toxic to humans, but sustaining for the immortals. Greek myth held that the sorceress Medea managed to kill Talos, the bronze giant who guarded Europa, the queen of Crete. Medea used a potion to put Talos to sleep, and then pulled out the nail (some call it a plug) that stoppered a vein running from his neck to his ankle. The ichor that kept Talos alive began to flow out of this one weak spot, and thus the god bled to death. Medea had a bit of a thing for spilling blood. She also murdered her brother, her husband's new fiancée, her sons, and her uncle.

But spilling blood in Greek mythology also gave rise to new life. Gaia, the goddess of earth, was the wife of Uranus, god of the skies. But Uranus had an unusual method of parenting: he forced his children to stay hidden in the earth (effectively inside Gaia's body). Fed up with this arrangement, Gaia offered a sickle to urge her children to lash out against the father who appeared not to love them. Cronus, the youngest and most ambitious of her sons, used the tool to castrate his father. From the bloodied testicles of Uranus, which fell into the sea, emerged Aphrodite, the goddess of love, beauty, and sexuality. Cronus's attack on his own father also spilled blood on the earth, and from those drops emerged other, more aggressive creatures, such as the Furies (also known as the Erinyes), who were deities of vengeance and whose eyes dripped with blood. Uranus's story shows two potent results of mixing blood with another substance. Mix it with the sea, and you create love. Mix it with the earth, and you generate vengeance.

Uranus survived the castration and prophesied that Cronus would be overthrown by his son. As a result, Cronus began killing his children, epitomized by the graphic and bloody *Saturn Devouring His Son*, a painting dating back to about 1820 by the Spaniard Francisco de Goya, who depicted Saturn (the Roman equivalent of the Greek god Cronus) eating his naked child.

One blood crime usually leads to another, in history and mythology. Literature and paintings serve to remind us that blood must not be spilled in vain. Blood is life, so it must present itself to us for a reason. And when blood runs, we look for consequences.

TO THIS DAY, WOMEN WHO DRAW BLOOD brutally loom larger in our imagination and generate more fears and myths than their male counterparts. We wring our hands with worry and ask how it could be possible that a woman could be brought to kill. I would answer that it's a miracle that women don't kill more than they do.

If we come to expect that a certain level of violence is inevitable in, say, a city of three million people, what makes murder by a woman any graver than homicide by a man? Are women supposed to nurture? Some of them. Aren't men supposed to nurture too? Yes, and many of us do. Neither men nor women have any business killing people, except in ritualized situations: executioner, perhaps, or soldier. Women go to war now too. They shoot, and they get killed. As we move to a more equal society, we find ourselves having to accept some uncomfortable facts: women will sometimes murder, and they will come home from war more and more often in body bags.

In the Biblical Book of Judith (which is in the Catholic Old Testament, but excluded by Jews and assigned by Protestants to the Apocrypha) a beautiful Jewish widow named Judith is upset that her people seem passive in the face of occupation by their Assyrian enemies. She gains the trust of the enemy general, Holofernes, slips into his tent one night, and while he lies drunk in bed — wanting her but never having touched her intimately — Judith beheads him. She then returns, with the decapitated head in her hands, to her people. The occupying army is spooked and flees, and the control of Israel returns to the Jews.

This scene inspired a painting by Caravaggio from 1598 to 1599, as well as two paintings, entitled *Judith Slaying Holofernes*, by Artemisia Gentileschi between 1611 and 1620. The first Italian woman to paint major images and stories related to history and religion,

Gentileschi depicts Judith in both works with her maid-servant, pulling back Holofernes' head to better cut it off while blood courses over the bed.

It appears that Gentileschi painted this shortly after she had been raped by a man named Agostino Tassi. Not quite eighteen at the time of the assault, Gentileschi was tortured during the course of Tassi's rape trial — the court subjected her to thumbscrews while demanding to know if she had been a virgin before meeting her attacker. Tassi was found guilty and banished, but he was not tortured or punished in any other way.

Women of the day were not permitted to paint unless they were married to painters. The year after the rape, Gentileschi married a minor painter and went on quickly to create *Judith Slaying Holofernes*. In addition to being a wife and mother, she became well known as a painter, eclipsing her husband in the public eye.

Some people have speculated that Gentileschi created *Judith Slaying Holofernes* to dramatize her own personal story of rape. Some have even suggested that the face of Holofernes is meant to bear a likeness to the rapist Tassi. I don't know if this is true. And I don't know if it matters. Artemisia Gentileschi was a painter. She drew upon her talent to create the image of a universal story: woman slays oppressor, thus liberating herself and her people. Gentileschi, regardless of her past, was doing what many great artists must do: retelling stories of oppression, bloodshed, and liberation.

Centuries later, in 1998, the French filmmaker Agnes

Merlet released the film *Artemisia*. Marketers originally claimed that the film was historically accurate, but
it depicted Gentileschi as having a bold romantic love
with Tassi — the man who was in fact convicted of raping
her — and this love was portrayed as the inspiration for
Gentileschi's art.

This turning of history on its head, and romanticiz-
ing of a relationship that began as rape and was recognized as such in court, led the feminist Gloria Steinem
and the art historian Mary Garrard (well known for
her work on Gentileschi) to circulate a critical flyer at
the New York premiere of the film. The flyer carried
the headline "Now that you've seen the film, meet the
real Artemisia Gentileschi." It continued: "In the film,
Artemisia Gentileschi and Agostino Tassi are presented
as voluntary and passionate lovers...In the fully documented trial of 1612, Agostino Tassi was charged with
and convicted of the rape of Artemisia Gentileschi.
Artemisia testified repeatedly under oath and torture
that she had been raped by Tassi. She described the event
in explicit and graphic detail, and her own resistance to
the point of wounding him with a knife...The theme
of the film is that Artemisia's sexual awakening, initi-
ated by Tassi, launched her artistic creativity...The idea
that a woman artist is the creation of a male mentor has
been a persistent myth in the history of art, frequently
asserted by artists and critics of the sixteenth and seventeenth centuries. So has the romanticization of violent
rape, as in the rape scenes in this film, and the idea that

women wish to be raped or fall in love with their rapists."

Artemisia Gentileschi overcame the trauma of rape and courtroom torture to become an independent, respected painter in her own time. Perhaps one of the most troubling aspects of the film is that it ran the risk of trivializing the life of Artemisia and the struggles she faced.

Metaphorically and perhaps literally, Gentileschi paid with her own blood in the bed of her rapist, and in the courtroom that slapped her rapist on the wrist but tortured her. In the seventeenth-century Italian court, it was said that Gentileschi scratched Tassi's face and penis in trying to fight him off. But in the film, the attacker is transformed into an artistic enabler. He inspires her genius. The film was criticized for trivializing the spilling, symbolically at least, of the blood of a woman who went on to become a great painter despite — not because of — her attacker. The red blood of humans is as sacred as the golden ichor of the Greek gods. To cause it to be spilled is serious business, and people do not take kindly to having these sacred matters trivialized.

ONE MIGHT IMAGINE, at first blush, that blood is something you're stuck with, like the thickness of your wrist or the shape of your head. You get what you get, and that's all there is to it. Perhaps subconsciously, we think of our blood as having fixed, inherent, unchangeable qualities that are entirely unlike the properties of our teeth, skin, or hair. Two years ago, to offer moral support to my brother, who was coping with the disease, I grew a "Movember" mustache to do my small part for fundraising and public awareness for prostate cancer. It was a pathetic mustache — seems that in my family, there were only so many facial-hair genes to be distributed among the siblings, and my older brother bagged all of them. Others might have grown in four days what it took me a month to produce, but it radically changed the look of my face. I was stunned by the change, only because I have never been a man to grow, discard, and then regrow mustaches or beards. But there are many ways that people adjust their looks on a day-to-day basis. We wear braces on our teeth; colour, perm, or restyle our hair; go to beaches or tanning salons; and engage in endless plastic surgeries to mask the effects of aging, to shed excess weight and make ourselves beautiful.

In *Truth and Beauty*, a memoir about her friendship with a writer who had facial deformities as a result of cancer during childhood, Ann Patchett wrote about how Lucy Grealy at first sought to correct the most profound of her deformities, but eventually fell victim to her own self-destructive obsession as she underwent operation after operation in the hope of making herself beautiful.

As I became transfixed by *Truth and Beauty*, I found myself rooting for Grealy in her early struggles to create a new face, but cringing when she appeared to indulge an increasingly dangerous vanity. I may have been too quick to weigh necessity and vanity, but I suspect that many others do the same. Perhaps this informs our own judgements about how we modify our own bodies — including our blood. Perhaps we are far too quick to qualify some interventions as right, and others as wrong.

As Patchett implies in her memoir, when we adjust our outward appearance by means of surgeries and injections, we are generally aiming to enhance our beauty. We believe that this will put us on the path to personal or social gain. Although we don't put it in the same category as plastic surgery, we also aim to enhance something about ourselves — our health, our strength, our endurance — when we change the nature of our blood. This reflects the notion of truth in Patchett's title. The truth is very much at issue when we play with our blood, because many adjustments to our blood are secret, illicit, and violate laws or social rules.

THE USE OF HUMAN EMBRYONIC STEM CELLS has generated years of controversy. Some consider it to be legitimate medical practice and scientific research; others say it violates human life or usurps a role best left to God. Like abortion, the creation of stem cells from human embryos — the early stirring of life after conception — incites heated debate about right and wrong. Before examining the issue, let's consider what stem cells are and why they are useful.

The bone marrow is the home of hematopoietic stem cells, immature cells that generate all other blood cells. If your stem cells don't work, or if they have been damaged as a result of radiation or chemotherapy, you may need a bone marrow transplant — carried out by means of blood transfusions — to stay alive. According to the National Institutes of Health, some medical conditions that might require such a transplant are leukemia, lymphoma, sickle-cell anemia, and severe immunodeficiency syndromes. Leukemia, by way of example, is a cancer of the blood or bone marrow, accompanied by an overgrowth of white blood cells (leukocytes) that crowd the bone marrow and impede it from functioning properly.

There are two ways to obtain donations for bone marrow transplants for people suffering from leukemia. The stem cells can be taken from the blood of another patient with a matching blood type. Or they can be harvested from the blood of the umbilical cord of a baby right after birth, and frozen until ready for use.

The donated stem cells enter the body of the patient by moving into the blood through a venous catheter, much like an ordinary blood transfusion. Miraculously, donated stem cells know how to navigate through the bloodstream and find their way into the bone marrow. It is a complicated and risky procedure. The donor's bone marrow, if not a good match, could perceive the body of the patient as foreign, and attack it. This is called graft-versus-host disease, and it can be fatal. Conversely, the patient's body might attack and destroy the donated bone marrow. This is called graft rejection. Infections are another serious risk. If the transplant is to succeed, the patient is likely to require antibiotics and multiple blood transfusions. The patient can feel acutely ill for weeks, and a full recovery can take six or more months. But by the time the process is over, the patient has a new system of blood and marrow.

Imagine a doctor going back in time and telling a bloodletter in 500 CE that one day, it would be possible to entirely replace a sick person's blood, as well as the spongy material inside big bones that generates the blood cells. Imagine telling Hippocrates that if a person's blood is failing, then it will be possible to replace it entirely. Consider his reaction if you informed him that blood originated from microscopic cells made inside the bones, and that by means of harvesting and then injecting stem cells, you could alter not just a person's blood but the life-giving stuff inside their bones. I like to think that Hippocrates would grow wide-eyed, grip the arms of his chair, bone up on life sciences, and apply to a modern medical school.

Bone marrow transplants are about replacing not only the blood but also what manufactures the blood, so that it will become free of disease. The first bone marrow transplants were conducted after the Americans dropped atomic bombs on Nagasaki and Hiroshima, effectively ending World War II. Thousands of Japanese citizens died, and thousands more suffered from the fallout of nuclear radiation. Scientists began to look into means of protecting people against the effects of irradiation. The first experiments were conducted on dogs and mice. In 1959, the first bone marrow transplants were attempted in humans, with little success. However, in 1969, the

American E. Donnall Thomas — who would later receive the Nobel Prize — demonstrated that it was possible to inject bone marrow cells into the bloodstream to build up the bone marrow and create new blood, and in so doing, to save the life of a person with cancer of the blood.

In the following decades, along with improvements in the science of bone marrow transplants, international registries were developed to assist with the complex process of matching patients to viable stem cell donors. Most bone marrow recipients will not have a family member with compatible blood, so they require a matching service to connect them to millions of potential international donors. Out of the ashes left by the most devastating bomb ever dropped has emerged not just the science of stem cell transplants but also a massive network of international cooperation to assist in locating blood donors for patients who might otherwise perish.

In modern society, although most people accept the need to intervene medically to save the lives of patients with blood cancers, a related issue has proven to be one of the most controversial aspects of contemporary medicine.

The controversy began in 1998, when U.S. researchers discovered human embryonic stem cells. The embryonic stem cells could be derived from the inner cell mass of an early human embryo called a blastocyst. About the size of a period at the end of a sentence, a blastocyst begins to form five days after fertilization in humans. Embryonic stem cells have an additional, nearly miraculous quality called pluripotency, which means that they can transform into any other type of cell or tissue in the body.

Through a process called differentiation, which means maturation, the embryonic stem cell can become a cardiac muscle cell, an islet cell to process insulin, a neuronal cell in the brain, or another cell in the body. But once it has started down the path of differentiation toward becoming, say, a cardiac muscle cell, it can never change course and become a neuronal cell. It has become too specialized, too committed to its future. The ability to control the direction of this maturation, to choose the sort of cell that we need and want them to become, is what excites scientists and makes them believe that the embryonic stem cell could well be tied to major medical advances that could help people overcome diseases and save many lives in the future. Indeed, *Science* magazine hailed the identification and culturing of embryonic stem cells as the "breakthrough of the year" in 1998.

Even former U.S. president George W. Bush, who introduced rules limiting federal funding of stem cell research in 2001, acknowledged at the time the potential importance of the science. On the day of his announcement of the funding restrictions, President Bush said: "Scientists believe further research using stem cells offers great promise that could help improve the lives of those who suffer from many terrible diseases — from juvenile diabetes to Alzheimer's, from Parkinson's to spinal cord injuries. And while scientists admit they are not yet certain, they believe stem cells derived from embryos have unique potential."

When the Nobel Prize–winning American physician E. Donnall Thomas was doing his work in advancing stem cell transplants, Canadians too were very involved with this pursuit. Ernest McCulloch, a Toronto-born physician and cellular biologist trained at the University of Toronto, teamed up with physicist James Till and proved the existence of stem cells in 1961. Two years later, they demonstrated two essential features of stem cells: they are capable of self-renewal and they can differentiate into more specialized cells. McCulloch and Till reached their conclusions by means of working with rats: first exposing them to potentially lethal amounts of radiation and then injecting them with bone marrow cells. They discovered that rats given more stem cells were more likely to survive. In their spleens, the surviving rats developed clumps of cloned cells that formed as a result of the stem cell injections.

Since that time, stem cell research has shot forward. One of the most notable steps took place in 1998, when the University of Wisconsin biologist James Thomson successfully removed stem cells from an early human embryo that had been donated for research purposes. Thomson's research ignited a debate that continues to this day: Are scientific and medical advances that could result from obtaining and using embryonic stem cells justifiable, in the face of moral and religious objections about the use of human embryos generally?

Those who oppose embryonic stem cell research say that the embryo—even a surplus embryo that is stored in an in-vitro clinic, just a few days old, and destined to be discarded after it is no longer needed by the parents who produced it—is a human life and must not be destroyed for the purposes of medical experimentation. The Center for Bioethics and Human Dignity at Trinity International University in Illinois cites passages from the Bible in arguing that human embryonic life exists for God's own purpose, not that of humans. Essentially, the center argues that people have no right to help themselves at the expense of other human life, and likens research using human embryos to other "horrific examples" of medical experimentation over the course of history.

On the other hand, many people—including President Bill Clinton, who first proposed federal funding for embryonic stem cell research in 1999, and President Barack Obama, who in 2009 relaxed restrictions on U.S. federal funding for embryonic stem cell research that had been imposed nearly a decade earlier by George W. Bush—believe that the potential medical and scientific benefits outweigh the moral objections. Even the late South Carolina senator Strom Thurmond—an arch-conservative whom I will mention in another context in chapter three—argued in favour of the potential benefits from embryonic stem cell research. Many people who support embryonic stem cell research make a key distinction: the embryo normally used in such research is just a few days old, has been cultivated in an in-vitro fertilization lab, would not be capable of developing into a human being without further medical steps, and would normally be discarded at any rate because it would not be used in efforts to establish a pregnancy. (In the case of in-vitro fertilization, many ova are fertilized with sperm but few of the resultant embryos are actually used.)

Although the debate has raged for years, other forms of human stem cells are available. Stem cells can be removed from umbilical cord blood without endangering newborn babies. And it is possible to use "adult" stem cells (which do not come only from adults — the term refers to the maturity of the cells, not to the age of the donor) that have already moved down the differentiation pathway described earlier. Additionally, adult stem cells can be harvested with little intervention and no harm: a skin biopsy, cheek swab, or blood draw can all provide adult stem cells. But adult stem cells do not have the pluripotency of embryonic stem cells. For many years, the debate was stalled on a key point: Should researchers destroy embryos to obtain more robust embryonic cells, or use the less promising adult stem cells, which can be harvested without ethical concerns?

Recently, we have seen a breakthrough that may satisfy both scientists and ethicists. In 2012, Shinya Yamanaka of Kyoto University won the Nobel Prize for demonstrating that adult cells could be returned to their embryonic state, only to have them develop into another type of stem cell. In a nutshell, he took an ordinary skin cell, reprogrammed it so that it resembled a primitive embryonic stem cell, and then reprogrammed it again along a new cellular pathway — without resorting to experimentation on embryos. In effect, he tricked a skin cell into becoming an embryonic stem cell. This heralded the creation of a new category of stem cells, known as induced pluripotent stem cells (iPSCs), which offer the benefits of embryonic stem cells without the ethical baggage.

More developments followed Yamanaka's discovery. In July 2013, the Japanese Ministry of Health, Labour and Welfare reviewed a proposal to use induced pluripotent stem cells in a clinical trial for age-related macular degeneration. It will take time to exploit all of the practical applications of Yamanaka's research, and the debate over the use of human embryos for stem cell research will not end tomorrow.

Reflecting in December 2012 on Yamanaka's Nobel Prize and on other advances in stem cell research, Matt Ridley made the following observation in the *Wall Street Journal*: "In the not-very-distant future, when something is going wrong in one of your organs, one treatment may be to create some stem cells from your body in the laboratory, turn them into cells of that organ... and then subject them to experimental treatments to see if something cures the problem." Given our newfound ability to create induced pluripotent stem cells, Ridley suggested the debate over embryonic stem cells might be fading into history. He wrote: "If stem cells derived from the patient's own blood are to offer the same therapeutic benefits as embryonic stem cells, without the immuno-logical complications of coming from another individual, then there would be no need to use cells derived from embryos."

The vigour and duration of the stem cell research debate underlines, once again, how deeply we humans react to issues involving the use and treatment of our own bodies and our blood. Blood stem cells come from the marrow of the big bones buried deep in our bodies. We call them immature cells, which is true because they are so young. However, the term *immature* is also ironic, to my way of thinking, because stem cells can also be thought of as the mothers of all other cells in the human blood and body. Mothers have great power and impor-tance, in our minds and evidently in our blood. Mothers produce the embryos that offer stem cells with such fan-tastic possibilities — embryos contemplated with equal intensity by researchers and ethicists.

SINCE KARL LANDSTEINER IDENTIFIED the main blood types in 1901, and since physicians began carrying out successful transfusions over the next years, the sharing of blood has offered opportunities to give in one of the most noble, selfless ways possible. When you give blood, you don't even get a grateful smile or hug from the recip-ient of your blood. Because your blood has been broken into parts, and there are many recipients, you will never know them, and they will never know you. You know that you are helping to preserve human life, and that general knowledge is good enough. The desire to reach out and help others in need reflects the best parts of our humanity.

Within minutes of hearing the news of the terrorist attacks in New York City and Washington on September 11, 2001, donors began lining up outside collection agencies in the United States. In Oklahoma City, for example, three hundred people were standing in line outside the Oklahoma Blood Institute by 11:30 a.m. on the day of the attacks. Nationwide, donors gave 1.5 million units of blood within two days of 9/11, creating a shortage of storage bags. The supply of blood eventually exceeded demand, and authorities asked donors to wait. After the Boston Marathon bombings in 2013, donors again turned out in such large numbers that they were asked to wait and schedule appointments to give blood later.

Since transfusions began, millions of people around the planet have received blood that helped save their lives. I am one of them, and I will be forever grateful. Today, according to the World Health Organization, people around the world donate more than eighty million units of blood annually. (A unit is about 450 millilitres, or just less than one American pint.) But the gift of blood has been a double-edged sword. It has brought out some of our worst fears and prejudices.

To set the politics of blood donation in context, I wish to go back to early- and mid-twentieth-century America, when racial segregation permeated virtually every corner of American life, and the mistreatment of blacks even spilled into the arena of health care. Take, for example, the Tuskegee Syphilis Study, which is now considered the most infamous biomedical research project in U.S. history. Doctors funded by the United States Health Service subjected hundreds of African-Americans to forty years of medical experiments without telling them that they had syphilis, without treating them for it, or intervening as they died and their wives contracted the disease and their children were born with congenital syphilis. The study began in 1932 and continued until 1972, when the media exposed it.

Another realm of health care in which American blacks felt the sting of rejection and disrespect relates to who was allowed to donate blood during World War II — when blood was in big demand and when the technology finally existed to rush large quantities of it to critically injured soldiers — and who was not. I will explore these politics — now almost three-quarters of a century removed — by remembering the life, words, and death of Charles Richard Drew.

Drew was born in Washington, D.C., in 1904, of black parents. His father was a carpet-layer and his mother a schoolteacher. He was a tall, light-skinned, athletic boy. He could have passed for white, had he been so inclined, but there was nothing about Charles Drew that sought to deny his heritage.

The society in which Drew grew up separated blacks from whites in schools, theatres, swimming pools, department stores, sporting events, and even in federal government cafeterias in the nation's capital. Drew attended the segregated Dunbar High School in Washington and then studied at Amherst College in Massachusetts.

Drew left the United States to attend medical school at McGill University in Montreal from 1928 to 1933, graduating second among the 137 students in his class. He went on to do two years of internship and a residency in internal medicine in Montreal. He obtained his M.D. and master of surgery degrees but had trouble finding a job in the United States. The Mayo Clinic, where he had hoped to work, turned Drew down, as did Columbia University. Allen O. Whipple, head of surgery at Columbia, gave Drew a frank explanation for his rejection.

Spencie Love, who wrote a biography of Drew in 1996, quotes a dean of the Howard medical school who witnessed the conversation and described it after Drew's death. Whipple told Drew that he was of the wrong race and economic class to treat the most wealthy and privileged of American citizens: "I am certain that you could do well with the average patient. But could you, with your background, feel at ease and render competent service if one of your patients in surgery were a Morgan, an Astor, a Vanderbilt, or a Harkness? We have such patients here. In the selection of our residents we choose only from among superior students and we take into account their family and personal background."

Drew taught for a few years at Howard University

and received an additional two years of training in surgery at Columbia at the hands of Dr. Whipple, the same man who had declined him earlier. Although blacks did not generally find openings for surgical residencies in the United States until the late 1940s and early 1950s, Drew, through a combination of persistence, personal charm, and his light skin colour, won over Dr. Whipple and gained a full training experience, including visiting hospital wards and treating patients.

The outbreak of World War II presented Drew, who at Columbia had become a leading expert on blood storage technology, with the opportunity of a lifetime. In 1940 — before the bombing of Pearl Harbor, which drew the Americans into the war — Drew was hired as the medical director of the Blood for Britain project, which was to ship liquid plasma from the United States to British soldiers who had been wounded in France. In 1941, he served for a short time as the medical director of the first American Red Cross blood bank, overseeing a pilot project involving the use of dried plasma for the anticipated wartime needs of American soldiers.

In November 1941, by which time Drew had returned to Washington, D.C., to lead Howard University's surgery program, the American Red Cross announced that it would exclude blacks from donating blood. Black leaders objected so vociferously that the Red Cross modified its policy, announcing in 1942 that "the American Red Cross, in agreement with the Army and Navy, is prepared to accept blood donations from colored as well as white persons ... In deference to the wishes of those from whom the plasma is being processed, the blood will be processed separately so that those receiving transfusions may be given blood of their own race."

Drew weighed in on the matter several times. In 1942, he said, "I feel that the recent ruling of the United States Army and Navy regarding the refusal of colored blood donors is an indefensible one from any point of view. As you know, there is no scientific basis for the separation of the bloods of different races except on the basis of the individual blood types or groups." Just six years before his death, while being honoured by the NAACP for his work in blood plasma research, Drew again criticized the blood segregation policies: "It is with something of sorrow today that I cannot give any hope that the separation of the blood will be discontinued ... One can say quite truthfully that on the battlefields nobody is very interested in where the plasma comes from when they are hurt. They get the first bottle they get their hands on. The blood is being sent from all parts of the world. It is unfortunate that such a worthwhile and scientific bit of work should have been hampered by such stupidity." Drew also commented on how the blood segregation policy was a "source of great damage to the morale of the Negro people, both civilian and military."

By the time of his accidental death at the age of forty-five, Charles Drew had risen to the position of head of surgery at Howard University College of Medicine. He might have continued to rise in prominence as a blood plasma expert, surgeon, university professor, and critic of discriminatory blood donation policies had he not fallen asleep while driving on a highway in North Carolina and crashed his car in 1950. After a long day of work at the hospital in D.C., Drew had been travelling with three other doctors to a medical conference in Tuskegee, Alabama. They had planned to help train black doctors and take part in a free medical clinic. The other passengers survived the accident, but Drew was thrown from the car and sustained fatal injuries.

For decades after his death, rumours persisted that the prominent surgeon and pioneer in blood plasma storage and shipping had bled to death because the Alamance General Hospital in North Carolina refused to treat him. But that is not true. Drew was rushed to Alamance General and treated promptly and thoroughly, as Love says in *One Blood: The Death and Resurrection of Charles R. Drew*. However, his injuries were massive, and his death could not be averted.

It's ironic that this tragedy should befall a man who had helped to save other lives by finding efficient ways to store blood plasma so it could be safely shipped from the United States to Britain in the early days of World War II. How symmetrical, and disturbing, that the surgeon who drew upon his medical training to challenge American policy barring African-Americans from donating their blood to white patients ended up being thrown from his vehicle and dying as the blood ran from his body.

It wasn't only black physicians who opposed the wartime blood segregation policy. At least one white physician, who would go on to become famous in his field, shared Drew's beliefs.

Bernard Lown, a cardiologist, creator of the direct current defibrillator, and internationally recognized peace activist, began studies in medicine at Johns Hopkins University in Baltimore in 1942, by which time the Americans had entered World War II. Lown, who was born in Lithuania in 1921 and moved to the United States at the age of thirteen, had family members who were murdered during the Holocaust, and he believed that the Allies were fighting in World War II to create a better world. During his time at Johns Hopkins, blood destined for patients in the university hospital was segregated into two categories: "colored" and "white." Because Lown knew that these were absurd distinctions, he took great delight in mixing up the tags on "colored" and "white" blood bags.

Lown was kicked out of medical school after it was discovered that he had supplied "colored" blood to a white patient from Georgia who had insisted that he not be given any "nigger blood." But colleagues raised a fuss on his behalf and Lown threatened to go public with the blood segregation policy; he was quickly reinstated.

Describing his acts of private rebellion at the Johns Hopkins medical school in more detail on his blog, Lown writes: "Black blood had to be kept apart from white blood. This was especially galling since apartheid in blood had no scientific basis. Yet it was being practiced in one of the leading medical schools in the country, an institution that prided itself on being a pioneer in promoting science-based medicine while it distinguished donated blood with tags labelled either C (for 'colored') or W (for 'white')... I decided not to partake in the immoral charade. Single-handedly I sabotaged the system. I did it with a black crayon. Whenever we were running low on white blood, I would take a number of bottles of black blood and add on the tag a mirror letter C to the one already there. The result resembled the letter W. Lo and behold, the blood was now white... For me it was both a medical and political maturing experience. I learned that if one wishes to effect social change, one must never walk alone, and that historical transformations are largely bottom up. Over my long life I have witnessed profound advances in diminishing the racist color line that pervades our country. This spurs a sense of unquenchable optimism. It affirms the poetic words of Martin Luther King Jr., that 'the arc of history is long but it bends toward justice.'"

Even in the time of the Second World War, American physicians understood that the body of a sick or wounded person receiving plasma or other blood products would not distinguish between the blood of a black or a white donor. The blood exclusion decision had nothing to do with science, and everything to do with society and politics.

WE LOVE OUR NEIGHBOUR so much that we are willing to give up our blood for a person, three streets or three thousand kilometres away, whom we will never meet or know. And yet we would much rather give, believing that we are healthy, than be put in a position of receiving. As blood transfusion science has revolutionized the ability of humans to give to each other, it has also raised our fears of each other. Could this donor make me ill? Might that blood bag be the cause of my death? We know that people who give their blood are likely to save someone's life — maybe our own — but we also fear that the same gift that moves straight into our own veins and arteries might carry a virus or disease that could kill or cripple us. One has only to look at the various exclusionary policies to see how concerned authorities are with preserving the safety of the blood supply.

Tainted blood, however, is an entirely different matter. In the 1980s, the tainted blood scandal that affected several countries entirely altered our collective psyche. Blood was no longer simply the pure gift that could save lives. In the minds of citizens across the world, it became a product that could kill. And kill it did. In Canada, some two thousand people acquired HIV and about thirty thousand people contracted hepatitis C after receiving tainted blood transfusions and tainted blood products to treat hemophilia. Some eight thousand people are expected to die as a result of having received bad blood in Canada over the course of a decade.

In *Factor 8*, named after a clotting factor given to

hemophiliacs, American documentary filmmaker Kelly Duda chronicles the blood harvesting scandal in an Arkansas prison and the blood's subsequent sale into Canada. Some of the tainted blood products entered the blood supply in Canada and other countries after Arkansas prison inmates — some of whom had unprotected sex in prison, shared needles for drug use, and contracted AIDS — were paid between $7 and $10 for each unit of blood they donated, which they collected themselves in inmate-run blood drives. The Arkansas Department of Correction used private organizations to sell the prisoners' blood at great profit to Connaught Laboratories in Canada, which used the imported plasma to create a blood-clotting factor later sold to the Canadian Red Cross Society.

The tainted blood scandal led to the creation of the Horace Krever's Commission of Inquiry on the Blood System in Canada, which reported its findings to the government of Canada in 1997. The Krever report noted that until Connaught Laboratories was informed in 1983 that the blood plasma it had been importing from the United States was tainted, the Canadian company "had not been aware of the fact that it had been processing plasma collected from prison inmates. The shipping papers accompanying the plasma had not revealed that the centre was located in a prison. They had simply referred to the source as the 'ADC Plasma Center, Grady, Arkansas,' without any indication that 'ADC' stood for 'Arkansas Department of Correction.'"

The Krever report recommended that new blood collection agencies be created in Canada to replace the Red Cross. As a result, Canadian Blood Services now collects blood in English Canada, and Héma-Québec collects it in Québec. The commission also recommended that the government compensate victims of the scandal. Federal and provincial governments in Canada have paid out billions of dollars of compensation, and civil litigation has been massive. Nearly $5 billion has been paid in compensation and to settle class-action lawsuits. Among its many other recommendations, the commission also said that, barring exceptional circumstances, donors of blood and plasma should not be paid, and that Canada needed far more stringent policies related to blood collection and blood safety.

Canada was not the only country affected by the tainted blood scandal. In what was surely the most catastrophic public health scandal of the century, people in the United States, France, the U.K., Ireland, Japan, and other countries were also contaminated with tainted blood. The global scandal spread just as the AIDS epidemic began to unfold in the world. Indeed, in 1981 — by which time tainted blood was already entering the market — HIV had not yet been identified as the virus that led to AIDS.

It took years to bring the contaminated blood supply under control, but one result was the implementation of a new rule in 1983 — in Canada and elsewhere around the same time — banning blood donations from gay men.

Experts tend to agree that it was a necessary step at the time. AIDS was highly prevalent in the gay community, and little was known about the disease, the virus that led to it, or how to contain it. Some three decades have passed since then, and science has evolved hugely, but the politics of blood donation have barely progressed.

In Canada, you cannot donate blood if you are a man who has had sex with another man (which classifies you as an "MSM"), unless you have been celibate for five years. This, ostensibly, is to prevent the HIV virus from entering the blood supply. However, each blood donation is tested for HIV, hepatitis C, and other viruses. There is a window of time—approximately two weeks—in which a donor might have acquired a virus, without that same virus showing up in a blood test. This is given as a reason for excluding MSM donations, even though heterosexual donors may also provide blood-carrying viruses that will not necessarily show up in tests.

The arbitrary, subjective nature of the rules barring or impeding blood donations from males who have had sex with males becomes very clear when one looks at the divergent policies from country to country. In Israel, France, Greece, and the United States, gay men are not allowed to donate blood. Canada recently eliminated its lifelong prohibition, and ruled that gay men who have been celibate for five years will be eligible to donate. New Zealand also sticks to five years. In the U.K., Sweden, and Japan, gay men can donate blood if they have been celibate for one year.

Gay rights activists and many others argue that the exclusionary rules are based on fear of homosexuality, rather than on science. They say that a promiscuous heterosexual who does not practise safe sex will pose a much greater risk for the transmission of HIV than a gay male who is faithful to one partner and who uses a condom.

Indeed, Spain and Italy do not bar blood donations from men who have had sex with men, but ask donors instead how many sexual partners they have had in the past six months. If the answer is one, they may donate. If it is more, then the donation is deferred. Mexico has also moved to allow gay men to donate blood.

Testing for viruses and pathogens in blood is far more sophisticated today than it was in 1983, when we did not yet have a test for the AIDS virus, and when the tainted blood scandal erupted in Canada and many other nations. We have more sophisticated means at our disposal for selecting donors judiciously: testing their blood, asking questions to eliminate those with risky behaviours, and so forth. But much like the decisions to prevent American blacks from donating blood destined for use by white U.S. troops during World War II. I would argue that the blanket ban on blood donations from sexually active gay men in Canada, the United States, France, the U.K., Japan, and many other countries is no longer based on science, but rather on lingering public bias that considers homosexuality inherently wrong and unsafe.

When it was announced in May 2013 that Canadian Blood Services and Héma-Québec would begin to accept donations from gay men who had been celibate for five years, nobody pretended that it would increase the quantity of blood donated, and nobody offered a rigorous scientific argument explaining the magic number of five years. It is time to stop creating rules that give credence to antiquated thinking about the inherent dangers of gay sex. A promiscuous heterosexual who does not practise safe sex is likely to pose more risks to the safe blood supply than a gay male in a long-term monogamous relationship. Federal health officials should give serious thought to a new policy that would take every reasonable step to ensure the safety of our blood supply, while not alienating and insulting potential gay donors. Rather than rejecting the safety of any blood donated by a gay male, people running blood clinics could give gays the same opportunities offered to heterosexuals. They should be asked clear questions about risky sexual behaviours. They should be screened out if their answers are unsatisfactory. Their blood should be tested, very carefully.

To refuse to allow blood donations from sexually active gay men has several negative consequences. It perpetuates stereotypes against homosexuality and robs the blood supply of vital donations. It runs the risk of discouraging heterosexuals who are sympathetic to gays from donating. It creates a system in which people who are desperate to donate might lie about their sexual orientation as a sort of act of political resistance. Indeed, in 2010 Canadian Blood Services won a lawsuit against a gay man named Kyle Freeman, who lied about his sexual orientation and donated blood several times between 1990 and 2002 as a protest against what he felt was an unfair exclusion. Although Freeman took blood donation rules into his own hands, others have opted for more concrete and open protests by staging campus demonstrations against blood drives.

Durhane Wong-Rieger, a former board member of Canadian Blood Services who has spent some twenty years working in areas related to blood policy, has decried the five-year deferral period. She says it perpetuates negative stereotypes about gay people and the safety of their blood — and has nothing to do with scientific evidence. From the standpoint of a recipient's health, the difference in risk between blood donated by a man who has had sex with a man and another donor is "absolutely infinitesimal," she said in a radio interview in May 2013. "The greater risk will be that someone who needs blood will not be able to have access to it," because of insufficient supply.

The American Red Cross could have taken a stand during World War II, by arguing that there was no reason to impede blacks from donating blood to white military personnel. Perhaps this would have helped the United States tackle serious problems of segregation and racial discrimination in an era when these issues were crippling the country. Today, federal officials in Canada, as well as Canadian Blood Services and the American Red Cross — where a lifetime deferral for men who have sex with men is still in effect — could show the same leadership with regard to blood from gay donors.

In June 2013, the American Medical Association (AMA) voted to oppose the ban by the U.S. Food and Drug Administration, which refuses blood donations from gay men. "The lifetime ban on blood donation for men who have sex with men is discriminatory and not based on sound science," AMA board member William Kobler said in a statement. The AMA argued that decisions to ban blood donations should be based on individual risk and not on sexual orientation alone. In July 2013, Arthur Caplan, who leads the Division of Medical Ethics at the New York University Langone Medical Center, also argued in favour of eliminating "an outdated, non-scientific regulation that bans [blood donations by] anyone who has had sex 'even once' with another man since 1977. Although many people died in the 1980s after they received blood donations that were infected with the HIV virus, much improved HIV testing has made the ban on gay blood donors obsolete," Caplan wrote. "The Food and Drug Administration acknowledges that HIV tests are highly accurate, with the risk from a unit of blood reduced to about one per 2-million units in the U.S. The worry is from the risk during the 'window period' which occurs very early after being infected with HIV when even current testing methods can't detect antibodies. But the Red Cross, America's Blood Centers and the AABB (a blood donation advocacy group) all support throwing out the ban."

Restrictions on who should donate should be based on science, on tests, and on meaningful questions designed to avoid donations from people who engage in risky behaviour. Even in an era of advanced medicine, when nucleic acid tests can reveal whether a donor has been exposed to HIV or hepatitis C before antibodies even show up in the blood, we rely on the honour system. When people give their blood so that others may regain their health, we still count on donors to tell us the truth about the most intimate parts of their lives.

In the world of blood donation, the safety of the blood supply depends on two key factors, which must be combined to maximize the benefits of synergy. We must use the best science in our laboratories, and we must ask many questions. As always, we will have to make wise decisions about blood donations. As for whether we deem it wise to accept the offer of blood, it will come down to three questions: How badly do we need it? What are the benefits? And what are the risks?

JUST AS WE COUNT ON the truthfulness and honour of blood donors, we also require it of the famous athletes whom we so revere, and from whom we draw hope and inspiration. Any superstar athlete who claims that his or her blood is clean, when it is not, runs the risk of creating a mighty scandal. At the intersection of honour and blood, we hold people to account. And so we should. If someone lies about blood being donated, people could die. If someone lies about racing clean on a bicycle or on the track, people will be shocked to the core. It is a matter fundamental to our sense of right and wrong.

Some eight years before he would be stripped of his victories and finally admit to Oprah Winfrey that he had repeatedly used performance-enhancing drugs and blood transfusions, Lance Armstrong was well en route to his unprecedented sixth consecutive victory in the Tour de France. On this particular day in July 2004, Armstrong and his teammates were riding in a bus on an isolated road after completing a mountain stage of the twenty-one-day race. One of the most gruelling sports events in the world, the Tour de France requires riders to race thousands of kilometres at high speeds, including several days climbing up and racing down the Alps and the Pyrenees. Needless to say, a race like that beats up your blood. Among other things, it drives down your natural red blood cell count, as well as your testosterone. Unless you resort to trickery, your body will deteriorate throughout the race — with some athletes deteriorating more than others. Cheating by such means as boosting the blood or adding to the testosterone count has become so widespread that to ride without chemical assistance is described, in the parlance of tour riders, as riding *pan y agua*, which is Spanish for "bread and water." Riding without chemical assistance is likened, thus, to toughing it out on the diet of a malnourished prison inmate.

On the day in question, Armstrong and his teammates pulled off a stunt so brazen that even teammate

Floyd Landis — who would win the Tour two years later, only to have his title stripped when he was found to have used testosterone — would admit later that he had never seen such a thing. The bus driver pulled over to the side of the road, feigning engine trouble. For an hour, every single rider on the team — Armstrong, Landis, and seven others — remained in the bus and underwent blood transfusions.

You cannot transfuse the blood of nine cyclists in a bus on a remote mountain road without meticulous planning involving numerous people, including athletes, coaches, doctor, and driver. Each cyclist needs to go somewhere — often travelling from one country to another and meeting secretly in a hotel room — to meet with a doctor who will withdraw about half a litre of his blood. The blood must be mixed with anticoagulant, preserved, and stored. It must be labelled carefully, so that each athlete's blood is kept distinct from that of the others. The blood must be refrigerated, with no electrical blackouts, thank you very much. (In 2003, one year before the group transfusion on the bus in the mountains of France, Lance Armstrong had to travel away from his apartment in Gerona, Spain, and worried about the possibility of a blackout in his absence. The blood might be compromised without Armstrong's even knowing it. The simplest solution was to hire a blood-sitter. Armstrong summoned Landis to his apartment to keep watch over blood bags stored in a refrigerator hidden in the master bedroom, checking the blood temperature daily to ensure that there had been no inconsistency in the supply of electricity. Landis came and provided the service, and later described the incident to the United States Anti-Doping Agency.) Machines must be purchased that can test the athlete's blood and see when it is ready for the transfusion — the receipt of his own blood back into his system. One machine, for example, monitors hemoglobin levels, and another is a centrifuge used to assess one's hematocrit (the percentage of red blood cells in the blood).

While keeping the blood cold, you need a courier to haul it past fans, journalists, television cameras, doping control experts, and others, and bring it to that bus taking riders down the slopes of a mountain. And then you need to hang those bags of blood above nine riders — one of whom is the most famous in the history of cycling, having survived testicular cancer to go on to obliterate his competition in the world's toughest cycling event for six years in a row, lying all the time about how he was riding clean. You need to hook it up, find a vein, and wait the hour or so it takes for the blood to drip into the riders' systems. With thousands of fans crowding each leg of the Tour de France, you need to dispose of the blood bags and other medical paraphernalia without anybody noticing, pretend that the bus driver has fixed the engine, and get rolling again so the riders can eat, rest, profess their innocence in the face of persistent questions from the media and doping control agents about whether they are riding clean, and race another day.

Athletes have always looked for ways to get a leg up on their competitors. In 1980, Rosie Ruiz pulled off one of the greatest hoaxes in modern sport by winning the women's category of the Boston Marathon in a record time of 2:31:56. This was about twenty-five minutes faster than a time she had earlier run to place eleventh in the New York City Marathon. In Boston, Ruiz finished the race three minutes ahead of the Canadian Jacqueline Gareau, the second woman to finish the race. American marathon legend Bill Rodgers noted that Ruiz looked unbelievably fresh at the Boston finish line. Ruiz was awarded the victory, but officials soon determined that she had taken the subway for part of the New York City marathon and had either taken the subway in Boston or found another way of getting herself close to the finish, so that she could run the last kilometre or so to "win" the world's most famous marathon. Ruiz was stripped of her title. Gareau was brought back to Boston for a proper ceremony as the rightful winner, and as the first Canadian woman to win the race.

As the scholar Mario Thevis says in his book *Mass Spectrometry in Sports Drug Testing*, people have been taking substances to gain an advantage in sport for thousands of years. Thevis notes that in the third to second century BCE, the Greek philosopher Philostratus observed athletes taking bread spiked with the juice of the poppy plant, which contains opium. Swimmers apparently used doping agents as early as 1865 during races in the Amsterdam canals, although it's beyond me why anyone would willingly swim in any Amsterdam canal in any century. Beginning in 1870, reports emerged about widespread abuse of narcotics, stimulants, and nitroglycerine by cyclists in six-day races. In the 1904 Olympic marathon held during the World's Fair in St. Louis, Missouri, the first man to cross the finish line was disqualified because he was found to have ridden in a car for part of the race. The second finisher was a British-born American runner by the name of Thomas Hicks. Hicks had run out of gas (pardon the pun) after ten miles (slightly more than a third of the way through the race) and had wanted to give up. His trainers pushed him on. They gave him two doses of strychnine, which is rat poison, but which in low doses also serves as a stimulant. He was also given raw egg white and a shot of brandy. Hicks had to be carried across the finish line and revived by doctors afterwards. Under current rules, Hicks would have been disqualified. But at the 1904 Olympics, he was awarded the gold medal.

Canada, of course, has its own infamous history of cheating on the running track. It's hard to imagine a single Canadian born before 1975 who does not know that in 1988, the sprinter Ben Johnson tested positive for anabolic steroids after winning the Olympic 100-metre dash in a record time of 9.79 seconds and trouncing his archrival, the American Carl Lewis, whom Johnson led from start to finish. To clarify, anabolic steroids do not alter blood composition per se, but they travel through the blood to enhance muscle strength and recovery.

Although Ben Johnson did not alter the composition of his blood, his blood carried steroids to the muscles. News that Johnson had cheated by using performance-enhancing drugs shocked Canadians just as profoundly as Americans and others were troubled by the truth — when it finally came out — about Lance Armstrong.

Johnson was turned into a pariah, and he endured a massive shaming in Canada. The same television networks that had replayed his victorious race over and over in slow motion camped outside Johnson's house and contributed, in a way, to his demonization in the public eye. I do not condone the use of performance-enhancing drugs by Ben Johnson or any other athlete, but it struck me then and it strikes me now that the man would not have attracted as much media attention if, instead of becoming a world-class athlete and cheating on the track, he had been convicted of murder. Having followed the world of track and field for years, and having been an entirely mediocre middle-distance and long-distance runner competing in dozens of races in high school, university, and afterwards, I remember being struck, and horrified, by suggestions from some quarters that Ben Johnson may have been too naive to know that he had been taking steroids. You don't progress through year after year of workouts, massages, and consultations with coaches and nutrition experts without becoming intimately acquainted with your own body and what is happening to it.

I never won an important track, cross-country, or road race in my life, but I could tell you what my resting pulse was on any given morning, without touching my wrist. I could guess, quite accurately, my own heart rate after a race, without having to hold two fingers up to my carotid artery, look at my wristwatch, and count for fifteen seconds. Athletes are aware of their blood, their heart rates, and what is going into their bodies. To my way of thinking, it was a slight to Ben Johnson's intelligence to suggest that he did not know what he was doing. Although Johnson initially issued vigorous denials, he had in fact been taking performance-enhancing drugs for the better part of a decade. He knew what he was doing, and he was fully aided by his doctor, Jamie Astaphan, and his coach, Charlie Francis, in his training and doping techniques. The truth came out later in the course of his testimony before the Dubin inquiry — a Canadian federal investigation into the use of drugs and banned practices in sport. Two years after Johnson was dethroned, the Dubin inquiry released its report. Not surprisingly, it documented widespread cheating among Canadian track athletes, particularly those who had worked with Francis. At the inquiry, Francis admitted to encouraging many of his athletes to take performance-enhancing drugs. The inquiry named eleven Canadian track athletes who had done so. (It is worth noting that many other Canadian athletes were competing clean.) Many other athletes came forth to offer testimony about what they had done to boost their own performances artificially. And many of the international athletes against whom Ben Johnson had been competing on world stages were also nabbed for taking performance-enhancing drugs, although few were brought down with as much ceremony, hand-wringing, and concentrated media attention as Ben Johnson.

Johnson, a black man, had a nervous stutter. He came as a child to Canada, from Jamaica. He did not do well in school. He lived with his devoted mother in a modest home in Scarborough, a suburb of Toronto. He had been a mighty Canadian hero when he won the Olympic gold medal, but after his shaming, some in Canada began to refer to him only as a Jamaican, disowning all ties with him, his accomplishments, and his failures. Johnson did not fare well in the aftermath. He apologized and said he would go clean, but he tested positive twice more in the intervening years for the use of steroids and a diuretic that can be used to mask the presence of other drugs in the body. Eventually, he was banned for life from competing on the track. Some described him as a national disgrace, and urged him insultingly to move back to Jamaica. At a charity event in 1998 in Charlottetown, Prince Edward Island, Johnson raced against a horse and a car. (After the 1936 Berlin Olympics, Jesse Owens also raced against horses.) Although Johnson's participation was voluntary, the pitting of a black man against an animal caused me to cringe. Among the most enduring and offensive racial stereotypes is that the black man lacks intelligence but has prodigious strength and sexual prowess. In Charlottetown, one decade after he ran a victory lap at the Seoul Olympics while Canadians cheered—for a few hours, until the music abruptly stopped—Ben Johnson had been equated with a horse.

The Dubin inquiry uncovered the fact that other elite Canadian track and field athletes had also taken performance-enhancing drugs. None, however, had climbed as sensationally high as Ben Johnson, and none had fallen so far from grace. The inquiry recommended more stringent procedures for testing in sport. In the more than twenty years since the inquiry's report was published, Canadian and international efforts to expose and punish cheating in sport have expanded exponentially, with Canada taking a leading role.

One of the similarities between the Ben Johnson and Lance Armstrong stories is how widespread cheating was among their contemporaries. Altering the chemical properties of their bodies and blood was not the sole province of Johnson and Armstrong. Many of their peers did the same thing. According to the United States Anti-Doping Agency, which hounded Lance Armstrong until it finally brought him down with the report that led to his being stripped of his multiple Tour de France victories, twenty of the twenty-one podium finishers in the Tour de France between 1999 and 2005 have been tied to likely doping through admissions, sanctions, public investigations, or hematocrit levels. In addition, of the forty-five Tour de France podium finishers from 1996 to 2010, thirty-six were by riders similarly tainted by doping.

Doping was so widespread that those practising it felt that it wasn't really cheating at all. In his book *The Secret*

Race, which documents his own doping history and that of his fellow Armstrong team members, Tyler Hamilton writes: "I've always said you could have hooked us up to the best lie detectors on the planet and asked us if we were cheating, and we'd have passed. Not because we were delusional — we knew we were breaking the rules — but because we didn't think of it as cheating. It felt fair to break the rules, because we knew others were too."

In the same book, Hamilton describes the mystery of having his own blood withdrawn so that it could be re-transfused later. "With the other stuff," he writes, referring to erythropoietin and testosterone, "you swallow a pill or put on a patch or get a tiny injection. But here you're watching a big clear plastic bag slowly fill up with your warm dark red blood. You never forget it."

After Hamilton's first blood withdrawal, for which he flew in Armstrong's private plane from France to Spain so that they could conduct the clandestine procedure without detection, he, Armstrong, and another teammate went out immediately for a bike ride along the Spanish coast. At the time, Hamilton was in the best shape of his life. He had just beaten Armstrong to win the prestigious Dauphiné Libéré bike race, including a gruelling stage victory up the notoriously steep Mont Ventoux in France. But on this day in Spain, just minutes after having their blood withdrawn, Tyler Hamilton and his teammates could barely ascend "a tiny pimple" of a hill. "We joked about it," Hamilton writes, "because that was all we could do. But it was unnerving. It shook me deeply: my strength wasn't really in my muscles; it was inside my blood, in those bags."

Indeed. As we learned from the Olympic debacle of triathlete Paula Findlay, whose anemia sucked the vitality right out of her on what was supposed to be one of the biggest days of her life, there is much power in blood. Athletes don't merely require killer quadriceps, reflexes like cats, and lungs like bellows. They also need great blood.

To succeed at the elite international level, and possibly make millions of dollars in prize money, appearance fees, and sponsorships, elite international athletes must pay close attention to their blood. There are ways to alter the composition of your blood legally, which is to say, ways that do not contravene the rules of the World Anti-Doping Agency, established in 1999 to promote the fight against doping in international sport.

One way, of course, is to train very hard. This will teach your blood to transfer oxygen more efficiently to cells in the muscles. Another way, which offers at best minimal gain, is to sleep in a well-sealed tent with a reduced supply of oxygen. This also teaches you to cope with less oxygen, thus making better use of what you have. Another technique, which offers the possibility of greater advantages than the oxygen tent, is to get on a plane and fly to an altitude camp — perhaps in New Mexico or Kenya — and to spend several weeks training there. Both techniques involve driving up your red blood cell count.

People who are in the business of measuring athletic progress in endurance sports will regularly encourage athletes to get their blood checked for hemoglobin mass (the actual amount of the red, oxygen-carrying cells in the blood) and hematocrit level. Exercise physiologist Trent Stellingwerff of Victoria, B.C., who specializes in advising athletes about optimal training techniques as well as legal physiological and dietary interventions, told me that his wife, Hilary Stellingwerff, tried to maximize the benefit of high altitude training by returning to sea level exactly fifteen days before her heat in the 1,500 metres at the 2012 London Olympic games. The extra blood cell count derived from training at altitude lasts only so long, he said, so they had to time the altitude work precisely. She made it through to the semifinal two days later, but missed qualifying for the final by just one spot.

Another legal way to expand the blood's capacity to work well under exercise is to train in heat. The stress of heat training leads the body to increase the amount of plasma in the blood. When you increase the plasma and total blood volume, you might be more successful at warding off dehydration and perform better in a race on a hot day.

Altitude tents, training at altitude, and training in the heat offer the opportunity to bring marginal improvements to one's red blood cells and plasma, Stellingwerff notes, but they do not offer nearly the advantage of a blood transfusion: "There is no comparison. The advantage of the transfusion is immediate and large (about a ten- to fifteen-percent increase), and you don't have to spend months at altitude to get it."

Prosecutors have been known to manipulate, suppress, or ignore evidence in order to obtain criminal convictions in court. Defendants sometimes lie. People cheat in business, every day of the year. Writers invent phony stories and label them as non-fiction, in order to attract more attention and readers. Teenagers lie; students, professors, and even school administrators plagiarize; presidents surreptitiously record the conversations of their political adversaries; and politicians of every stripe fabricate stories to protect themselves. Deception permeates human society. Athletes — especially those of the elite variety, who are in a position to earn huge sums of money — are no different. Some will be honest, and others will not.

Testing for performance-enhancing drugs and doping is improving every day. The World Anti-Doping Agency is now making use of a biological passport, which provides a sort of molecular and physiological blueprint of an athlete's blood values over time, enabling experts to arrive at conclusions about whether any given athlete has cheated, regardless of whether they catch a banned substance in their blood or urine. Using ever more sophisticated technology, testers will continue to chase athletes, some of whom are indeed cheating but will manage to avoid detection. But the athletes and their coaches and medical advisors are sophisticated and strategic too. To date, they have been the ones to develop ingenious ways to cheat. Sometimes, the testers catch up faster than the cheaters anticipate.

Here is one stunning example of magisterial deception. When Lance Armstrong had banned substances in his body and learned that a drug tester was setting up shop in his hotel lobby, one of his doctors slipped past the testers, retrieved a saline solution, stuffed it inside his jacket, and snuck it back into the hotel. In the privacy of Armstrong's room, and before the tester was ready, the saline solution was hooked up to the cyclist and allowed to drip into his blood. This drove down the hematocrit in Armstrong's blood — changing the percentage of red blood cells in relation to plasma — thus making the cheating impossible to detect when Armstrong was required later that same day to submit a test. For years, Armstrong used the argument that he had never tested positive — not completely true, by the way — to protest his innocence and deride those who didn't believe he was riding clean. But the only reason he didn't get busted for testing positive was because he and his advisors managed — for a time, and only for a certain time — to stay one step ahead of the testers.

In the end, Armstrong was dethroned by many bits of evidence that had little to do with the accuracy of laboratory tests. For example, fellow cyclists Tyler Hamilton, Floyd Landis, and George Hincapie testified under oath about the doping they had done and seen Armstrong do. They testified, as well, that he not only took drugs and doctored his blood but compelled cyclists racing for him — cycling is a team sport, with one leader supported by a cast of "domestiques" — to do the same. But Armstrong was also undone by the testing laboratories. Years after his final Tour de France victory, in 2004, testers were able to go back and review his old samples again, using more advanced technologies than had been available when the samples were taken. In this way, they found incontrovertible evidence that Armstrong had indeed used performance-enhancing drugs and banned processes. When the truth could not be suppressed, Armstrong came clean on television, in two back-to-back confessions to Oprah Winfrey.

It is too soon to say what will become of Lance Armstrong. It had once been rumoured that he would enter politics, but it is hard at this time to imagine him persuading voters that he is a credible candidate for any public office. More interesting, to me, is the future of international sport. We have seen many ugly incidents in the recent history of sport: the Tour de France and its multiple drug scandals; the dethroning of Ben Johnson. It should be noted that we are not just talking about men. In the 1980s, there was such rampant cheating in sprints and field events in women's track and field that many of the records set back then remain completely out of reach by today's most successful women athletes. Contemporary female athletes are tested far more rigorously than were their counterparts three decades ago, and are able to get away with far less use of testosterone, steroids, and other aids.

Will we one day give up, and agree that cheating is a fact of life? Will we simply allow the best-conditioned athletes to use the best cocktails of blood, drugs, and genetic alterations and see who ends up on top? I doubt it, and I hope not. We won't allow our athletes to cheat with impunity any sooner than tax authorities allow offenders to lie on their tax returns, or voters allow liars to return to public office.

Will we convince more athletes of the health risks associated with performance-enhancing drugs and blood transfusions? Some of the related dangers include liver damage, blood as thick as sludge, heart damage, and heart attacks. It is imperative to educate young athletes, especially, about the danger of playing with drugs to enhance their performance. We can't let those efforts flag. But let's be honest. Most people who cheat know what they are doing. And to hell with the consequences, especially if they win.

As long as the benefits associated with success in sports are substantial, and the risk of getting caught appears minimal, I suspect that the ways some people pursue changes to their blood composition will be no different than the myriad other ways that some people try to weasel ahead in every other arena of life.

BLOOD, IN ALL OF ITS IMMENSE COMPLEXITY, has become merely another field in which we can try to gain an advantage. If we can gain this advantage by playing within the rules of the day, we are considered to be of moral character. We are unlikely to lose our medals, be banned from competition, or be shamed publicly. Now that we know just how vital our oxygen-carrying hemoglobin is, and how necessary water is in our plasma, we will find it increasingly easy to boost the blood or manipulate a gene. Those who believe in fair play — not just as role modelling for children, but as an approach to living well at any age — will continue to feel hurt and betrayed by the deception that some will employ to nab an Olympic gold medal or pocket all of the money and fame that comes with winning the Tour de France. We will continue to be astounded by the ingenuity of those who behave as if the end justifies the means, and we will be comforted and encouraged by detectives of the blood, who will occasionally not just catch an individual scammer but bring down entire networks of deception. The audiences — in the Olympic stadium, or seated before their own televisions and at their computers — will cheer loudly and vociferously for the noble winner, but they will jeer, prod, and mount a ceaseless moral attack against any hero who is brought down for having manipulated his or her blood. Don't believe me? Just ask Ben Johnson and Lance Armstrong. They played with their blood. And they were found out. And they were ruined.

There are drug testers and there are those who decide to strip former winners of their medals. But there is also the court of public opinion. In that court, if you alter your blood to win on the track, or on your bike, you have made a pact with the devil. To win a medal and a few million dollars, you allowed your morality — your very blood — to become impure. The devil will dance with you, and when your time is up, he will dance on your grave. But what about the friends, the lovers, the children, the former elementary school teachers, and all the unnamed armchair athletes who cheered for you as you transcended the challenges of cancer or poverty to inspire us with the magic of your strength and your resistance to pain? Their judgements will last forever, and in their eyes, there is only one observation. Once you were pure, and then you were not.

The world of sport serves as the perfect mirror. We look into it and we see our deepest values either respected or trashed. Blood, to us, is sacred. We offered it to the gods for thousands of years. In blood, we see the brightness and the movement of humanity. We open up a hole in our arms and give it away to extend the life of someone we will never meet. There is hardly a substance we hold more dear, or consider more valuable, than our blood. This is how much blood matters: even non-believers treat it as sacred. We want our sacrificial gestures, our gifts, our artists, our scientists, and our athletes to respect blood. In a world where we fret that nobody seems to care about anything, blood still counts. And we want it clean. As clean as spring water.

THREE

COMES BY IT HONESTLY:
BLOOD AND BELONGING

IN 1979 — JUST A YEAR OR TWO before stories about HIV and AIDS began to appear in the media — I travelled as a volunteer with Crossroads International to spend the summer working in Niger. The country has a population of about sixteen million people. It is the largest country in West Africa, consisting mostly of the Sahara Desert and including a small, more densely populated, arable strip in the south of the country. Niger often enjoys the dubious distinction of being named one of the poorest countries in the world. It was my first trip to sub-Saharan Africa. I was twenty-two years old. I thought I had embarked on the adventure of a lifetime, but did not expect that the journey would be an introspective one.

I was travelling with six other Crossroads volunteers — all white francophones from Québec. I liked and trusted them, and some of them have remained good friends for more than three decades. However, as soon as the plane landed in Niamey, Niger's capital city, and I stepped down onto the tarmac, I felt not just an oven of heat but also an explosion of unanticipated emotion. My very molecules, it seemed, screamed with desire to connect with the people of Niger. I longed for their acceptance, and for their recognition of my own ancestral history. Through my father's family, my ancestry dates back to the abduction of millions of Africans who were shackled and shipped against their will to be enslaved in the United States. I had grown up in a mixed-race family in a white suburb. By the time I was twenty-two, I had been searching for years to cement my own growing sense of black identity, and this was my first opportunity to travel meaningfully in Africa.

In the Americas, if your black ancestry dates back to Africa prior to the transatlantic slave trade, you can't return to connect with your distant relatives in the same way that a Polish-Canadian might go to Kraków, or an American or British citizen of Chinese descent to Shanghai. You have a general idea of who your people are, but this is made abstract by the diversity of the continent. Africa is the second-biggest continent in the world. It has fifty-four countries and one billion people speaking countless languages, celebrating many cultures and religions, and embracing a range of lifestyles. Africa is so big and complex that it is no more meaningful or helpful to generalize about it than it is to generalize about the entire planet. So if you happen to be a black Canadian, American, or Jamaican whose people came to the Americas via slavery, chances are next to zero that you will track down the exact villages of your distant African ancestors.

Alex Haley became one of the world's most famous writers for appearing to overcome these barriers in his novel *Roots*. First published in 1976, *Roots* purported to trace Haley's African-American family history all the way back to a man named Kunte Kinte, who had been abducted from the Gambia and sold into slavery in the United States. *Roots* sold millions of copies, was adapted for a major television miniseries, and encouraged a tidal wave of zeal among African-Americans and others who felt inspired to track down their genealogies.

Later, Haley paid to settle a lawsuit having to do with plagiarizing the work of another novelist, and he was roundly criticized for having used an unreliable griot, whose role is to memorize and recite many generations' worth of family and village history. It is possible that the griot was bribed to drum up a convenient story about Haley's family connection to Africa. In fact, Haley was so criticized on these fronts that, although his novel stands out as one of the most famous twentieth-century African-American literary texts, it remains excluded from the prestigious *Norton Anthology of African American Literature*.

Most people I have met who read *Roots* don't appear to care that the family history was somewhat fictional. It was a novel, after all. *Roots* still stands out as *the* twentieth-century literary celebration of the deep ties between America and Africa. It certainly had burrowed deep into my own soul by the time I found myself exploring the streets of Niamey.

In my first days there, while we stayed at a centre for young people and waited for the signal to begin a rural tree-planting mission, I took every opportunity to slip away from my fellow volunteers from Québec, and to meet one-on-one with the people of Niger. Every morning, I ventured to a street corner to drink the coffee made by a man named Moussa. He was young and friendly, and asked me a million questions, and taught me a few phrases of Djerma, his language. I spent several mornings in Moussa's company, sipping a bizarre mixture that he called "café" but which contained instant coffee powder, a tea bag, sickly sweet and syrupy condensed milk, and water that had been heated over a three-stick fire. I kept returning, anxious to learn more and more Djerma and to test out a series of salutations on Moussa and his other customers, until suddenly I could return no longer.

The last thing I remember eating, before becoming sick, was green olives, which some kind person had served to me in his modest home. I suspected that bad water caused my illness, but since the green olives were the last thing I'd eaten, I would not eat them again for decades.

At any rate, I became violently ill, with a high fever and bouts of vomiting and diarrhea. Soon, I could no longer walk or stand. The illness raged on. It is possible that my friends from Québec saved my life when they lifted me into a taxi and took me to the hospital in

Niamey. It was crowded, but the staff made room for me. They found me a bed, but as for food and drink—when I would be well enough to have any—it would be up to my friends to bring it to me. This was the case for all other patients too. If you had to be hospitalized, you needed someone to bring you food.

I remember a doctor saying that I had gastroenteritis. I was so dehydrated that I required an intravenous drip. My blood was tested. The doctor said that my red blood cell count had fallen dangerously low and that I needed blood transfusions. As I now know, the adult body contains about six litres of blood. In the simplest terms, it consists of plasma (which is mostly water), white blood cells for coping with infections, and red blood cells for transporting oxygen. The oxygen transport function is so important that, as we saw in chapter two, elite athletes in endurance events experiment with various means to boost their red blood cell count.

Many Africans die needlessly of gastroenteritis because of lack of clean water and advanced, affordable medical treatment. My friends and I did not hesitate: we accepted the advice of the doctor. It was 1979. Nobody had heard of HIV/AIDS yet. (I was lucky to have become ill before the disease turned into an epidemic, taking thousands of lives and affecting blood safety around the world.)

I have always had a mild fear of needles. Before developing diabetes in mid-life, I dealt with it by turning my head away when I had blood withdrawn for tests. (Now I have to look when I inject myself with insulin.) But that summer of 1979 in Niger, as I dropped from about 150 pounds to a skinny 125, and even as fever, nausea, and pain racked my body, I could not stop thinking and worrying about those bags of blood that took all of an eternity to drip into me. I wanted it over and done with in the time it takes to insert and withdraw a needle, but instead I lived with that needle in my arm, with those intravenous bags, and with that blood for hours. At first, I tried not to look but couldn't avert my eyes perpetually. Next, I stared at the hanging blood bag, which was shrinking by reluctant degrees, in an effort to overcome my fears. It didn't work. I imagined the person or persons who had donated the blood. African? European? North American? How far had the blood travelled to arrive in the Niamey hospital? Had it been flown or trucked in a five-hundred-millilitre bag in a cooler packed with ice? In my state of illness and anxiety, I hoped that they had accurately tested the blood of the donor so that it would match my own: A-positive.

Now that I was on a hospital bed with blood to absorb, I no longer felt preoccupied by the idea of having my own heritage — my own blood, or so it felt — accepted by the people of Niger. It didn't matter in the slightest whether I received blood from an African, an Asian, a European, or an American. What mattered was that our blood types matched. What mattered was that someone — living near or far, I will never know — had provided blood so that I might go on living.

My friends from Québec took care of me every day,

bringing food and water and sleeping on a mattress by my bed. It became instantly possible for me to love and accept them, while simultaneously wanting to discover Niger and its people. One love did not preclude the other. One aspect of my own heritage did not rule out the other. I made a promise to myself: that when I recovered and left the hospital, I would never worry again about how people imagined or interpreted the nature of my blood. Why should I worry about what others might think? I knew who I was, and I knew my family background, and I no longer felt any need to prove or establish it in public. This became the gift of my illness, and of the donated blood that helped me recover.

The experience of falling ill and receiving blood changed my emotional makeup and relaxed my self-concept. If the donor had been a black man or woman, did the transfusion make me more African? Of course not. My blood had been boosted, and changed, but I was the same person. Its real impact, in terms of my body and soul, was an entirely private matter. Race has nothing to do with one's true blood, or skin colour, and everything to do with perception — self-perception, and the perceptions of others. I finally came to this understanding in the hospital room in Niamey.

As far as the body went, the physical changes brought about by the transfusion were locked in my blood-stream. Assuming that the medical diagnosis was correct, the donated blood restored my red blood cell count. I regained my health within a week.

Thanks to a transfusion that helped me avoid dying at a young age — before falling truly in love, raising children, writing a book, or even understanding the meaning of my first volunteer stint in Africa — my soul shifted. It changed weight. I felt the weight of identity preoccupation lift off my shoulders. I no longer cared who saw what in my ancestry, when they looked into my eyes. I was both black and white, and no longer needed affirmation from the people of Niamey. It's as if a higher power had been looking down at me and said, "Hey, that coat looks heavy. May I hang it up for you?" The transformation of my blood had its most significant impact not on my body, but on my way of thinking. The confluence of my blood and that of another human was like two rivers meeting. One set of corpuscles merging with another kept me alive, and in the truest sense of the word, made my own blood truly mixed.

In this chapter, I will explore the meaning of blood, particularly in light of how it defines us in our private and public lives. I will look at how notions of blood affect the relationships we create with our families, careers, and nations. Situations of racial ambiguity tend to expose our unexamined assumptions about blood and belonging, so I will meditate on topics that have held my attention for years: how governments, courts, and social groups have navigated through disagreements on matters of black, white, Asian, indigenous, and national identity.

Blood has entered our minds as inextricably linked to personality and destiny because our earliest, foremost

Western thinkers suggested that was so. It's an awfully seductive fluid. When it leaves the body, it's a big deal. People might die. People might be accused of attempted murder, or worse. Even when it is supposed to spill — think of menstrual blood, for example, or of the blood from the broken hymen of a virgin on a wedding night that unfolds according to all the ancient rules — it has power and significance. Maybe it is impure. Maybe it could damage you. Maybe that menstrual blood could spoil food or rob a man of his hunting power. Or maybe it is the blood of the virgin, suggestive of innocence and perfection. In addition, blood acquires holy significance in the world's pre-eminent religions. Christians consider Christ's blood to be sacred, and imagine that they drink of it when they lift the holy cup of wine to their lips. Judaism and Islam have intricate rules about how animals are to be bled and how blood must be absent from food.

Blood filters into our consciousness in ways that surpass any other bodily fluid or any bone or tissue. It has become such a powerful metaphor for personality that we have forgotten that it is an idea — not a reality. It helps us imagine ourselves. But perhaps it helps us too much. We have bought the metaphor so fully that we have come to believe it to be fundamentally true.

MY FATHER (A HUMAN RIGHTS ACTIVIST) and older brother (a singer-songwriter), both named Daniel (or Dan) Hill, were known across Canada in their respective fields and had each written books long before I wrote novels. By

the time I was in my late teens, Dan's songs were playing on radio stations across Canada and around the
world. With his long hair, romantic eyes, soulful voice,
bare feet (on stage, at least), arm wrapped over guitar,
and gut-wrenching lyrics of heartbreak that a friend of
mine teasingly called "songs by which to slit your wrists,"
Dan's voice permeated cafés, the back seats of cars, and
wedding halls.

I adored my brother, and still do, but his fame added
to my motivation to leave Toronto after high school
and forge my own identity somewhere else. We met up
here and there to play squash, go for a run, or share a
meal, but even when oceans separated us, Dan followed
me. On the southern Atlantic coast of Spain, in a village called Sanlúcar de Barrameda, where I had gone to
write and where the only non-Spaniards apart from my
first wife and me were two lonely Mormon missionar-
ies who seemed wanted and loved by nobody, I heard
his big hit "Sometimes When We Touch." I heard Dan's
song in a taxi in Yaoundé, the capital of Cameroon,
where I had gone to work as a volunteer with Crossroads
International. Perhaps the most haunting reminder of
his omnipresence reached me in Québec City, where I
had gone to study at Laval University. Not yet fluent in
French, I'd been in the city for only two days and was
hunting for an apartment to rent. I was having a hard
time of it. Prospective landlords had been hanging up on
me when I phoned them to ask about their apartments,
and I had taken to the streets to scour for "*à louer*" signs

in the windows of houses and apartment buildings. One man had just made me an offer that, I suppose, he felt that an octoroon — this is what he called me ("octoroon" being an antiquated Southern term for a person who is supposed to have one-eighth black blood) — could not refuse: free rent in exchange for certain "nocturnal services." I kept on walking and soon found myself hungry and despondent in the Carré d'Youville, near the gates to the Old City. Cars and taxis buzzed about, and pedestrians crowded the tourist shops. But then I walked past a theatre for the performing arts, and there on the massive advertising billboard was my brother's name and giant face. Figuratively and literally, I felt that Dan's successes were spiking my own blood pressure.

Later that day, I found a bachelor apartment to rent. I stayed in Québec for two years, travelled a few times to Africa and Europe, and worked in Winnipeg and Ottawa. With one brief exception, I did not return to live in Toronto until eleven years after finishing high school.

Years later, my novels and non-fiction books began to find a few readers. One never knows who will show up, or not turn up, at a public reading. At my first literary reading in Hamilton, Ontario, four people attended — the event organizer, the bookseller, the bookseller's assistant, and me. Committed writers just take it and carry on, so I did too. When people heard that I was a writer and knew of my family connections, they would often say: "You come by it honestly" or "It's in your blood!" If writing is in my blood, my circulation is awfully slow. I had been writing since my childhood, but had no impression that a knack for writing flowed in my veins. To me, it felt like twenty years of effort finally paid off when *Some Great Thing*, my first novel, was published in 1992. Eight books later, I still don't feel that writing is in my blood. It is in my brain, and in my work, and in the hours I have invested and the hours I have yet to invest in the development of my craft.

People often ask me if any of my five children will become writers. I don't know. All I really wish is that my children lead rich, fulfilling lives. My own worth and identity are not validated if my children become writers, or negated if the children opt to drive trucks or fly planes. If they do become writers, will it be because the writing gene in my own veins slipped into theirs? I prefer to imagine that it will be because they worked bloody hard at something that they felt was an achievable, ordinary goal. I like to think that they heard me typing late into the night a few thousand times, and that the process of pitching one's soul onto the page seemed normal to them. If it was part of my daily life, they might think it could be part of their lives too.

Parental expectations can crush the spirits of children. I sometimes jokingly say that as the son of immigrants to Canada, I understand full well that the last thing many immigrants want is to see their son or daughter become a novelist. They are looking for doctors, lawyers, engineers, architects — any professional career that they hope will protect their children from economic and social vicissitudes visited on earlier generations in the family, in other parts of the world. The expectations can be suffocating when expressed along lines of blood.

I want to share one episode from my childhood. My father loved my brother, sister, and me very much. We were lucky to have him, and we loved him too — especially after we hit the magic age of seventeen or eighteen and moved out of the house, which made us free to engage with him on our own terms, and freed him of trying to force-feed professional career ambitions down our throats.

I credit my father's personality and energy for infecting me with an enthusiasm for life and for writing. But when we were children, Dad never spoke of his own failures, and chose instead to hammer home all of the accomplishments of his ancestors. For example, my great-grandfather, Daniel Hill I, had been born in Maryland in 1860, just a step outside of slavery. He graduated from Lincoln University and went on to become a minister of the African Methodist Episcopal (AME) Church. His son (my grandfather), Daniel Hill II, graduated as well from Lincoln, went on to complete graduate studies, and became an AME church minister. Following in this tradition, my father earned a doctorate degree and playfully but frequently drove home that he had a "Phud" (Ph.D.) from the University of Toronto. He perceived that ambition and professional success were in the family blood.

My father was also fond of citing the accomplishments of our Flateau cousins. He had an older sister,

Jeanne Flateau, who lived in Brooklyn with her husband and their seven children. Jeanne had raised the children and managed to keep up her own career as a social worker. I felt close to those cousins, and still do, but as a child I hated it when Dad rhymed off all of their successes. I felt that he was telling us that we had a super-intelligent, high-achieving strain of blood in one branch of the family, and implying that we — the Hill children in Toronto — were of a lesser branch. The Flateaus, Dad seemed to be saying, were in the family arteries. We were in the capillaries. It was particularly confounding because Dad let us know that he did not expect all that much from us, but that anything less than perfection would be unacceptable. This made me crazy.

One day, while attending Grade 9 at the University of Toronto Schools (a private high school), I came home, swallowed nervously, and told my father that the very next day I expected to fail a biology test.

He looked at me firmly and said, without the slightest hint of irony or playfulness, "Hill kids don't fail."

I did fail the exam, and felt all the worse as a result of what he had told me: that this failure violated a rule of my family blood.

HOW DID WE COME TO EQUATE blood with the transmission not only of skill and talent but the right to lead and inspire others? How is it that we have bought into the idea so fully that for the longest time we believed that bluebloods had a moral right to rule over others?

Part of the reason may reside in the physicality of blood; the way it moves, circulating through the body, being sent out and returning—unless it is spilled in the drama of an accident, or is made to spill as the result of a purposeful attack. But as we saw in chapter one, we have also held on to some part of the concepts that Hippocrates developed in 400 BCE, when he theorized that personality and health were determined by the presence and balance of the four humours.

In *The Nature of Man*, Hippocrates wrote: "The body of man has in itself blood, phlegm, yellow bile and black bile; these make up the nature of his body, and through these he feels pain or enjoys health. Now he enjoys the most perfect health when these elements are duly proportioned to another in respect of compounding, power and bulk, and when they are perfectly mingled."

Six hundred years later, the Roman physician and philosopher Galen argued that the preponderance of one particular humour determined a person's basic personality type. Blood was thought to accelerate the spirit, and thus we have come to employ the word *sanguine* to describe fundamentally optimistic people. If you are a positive person, Galen led us to believe, it's because it is in your blood. It's not a huge leap to assume that talent, and God-given rights to lead or inspire, might also reside in the blood.

Because of the public successes of my father and brother, I've always been intrigued by the lives of people who pursued the same public careers as their famous parents or siblings. When singer-songwriters Julian Lennon and Adam Cohen began to attract attention, I felt that I understood what it might mean to them to have grown up in the shadows of John Lennon and Leonard Cohen, respectively. Their careers might forever be situated in their fathers' shadows.

In March 2013, the young Canadian singer-songwriter Zoe Sky Jordan gave an interview on CBC Radio about the release of her new album, *Restless, Unfocused*. The interviewer noted that Sky Jordan is the daughter of musicians Amy Sky and Marc Jordan, both of whom have had successful careers performing their own work and writing songs for other recording stars. Early in the interview, Sky mentioned that in her high school years, she had resisted following in her parents' footsteps, choosing instead to explore other art forms, such as pottery and dance. However, after graduating, she did not feel the call to go to university and chose instead to launch her own career in music.

"People, especially in Toronto, are always coming up to me and talking about my blood," she said in the interview.

Just as music lovers pay attention to family ties among recording stars, voters follow family dynasties in politics. There were U.S. senators Bobby and Ted Kennedy, brothers of President John F. Kennedy. Cuban president Raúl Castro is the younger brother of the famous revolutionary and the country's former, long-time leader Fidel Castro. Indira Gandhi, the third prime minister of India, was the daughter of her country's first prime minister, Jawaharlal Nehru. The current leader of North Korea, Kim Jong-un, is not just the son but also the grandson of his country's previous leaders. Since the founding of the North Korean state in 1948, Kim Jong-un, his father, Kim Jong-il, and grandfather Kim Il-sung have been the only leaders. As North Korea escalated its bellicose anti-Western rhetoric in 2012 and early 2013, some Western observers wondered if the newly appointed supreme commander of the Korean People's Army was trying to climb out from under the weight of his family dynasty and create his own identity as North Korea's new leader.

When Justin Trudeau was elected to serve as a Liberal member of Parliament in Canada, and again when he won the leadership of the Liberal Party in April 2013, I could not help but wonder how he would hold up when constantly compared to his father. The late Pierre Elliott Trudeau, the flamboyant political superstar who was adored by some electors and detested by others, served as Canada's fifteenth prime minister from 1968 to 1979 and again from 1980 to 1984. In the barrage of news coverage about Justin's march toward the leadership of the Liberal Party, it became virtually impossible to read or hear any detailed news report that did not mention his father, and many speculated about the idea that politics ran in the younger Trudeau's blood.

Perhaps some part of the contemporary voter's soul longs to be ruled by dynasties. Even in anti-monarchist countries such as the United States, dynasties (think of presidents John Adams and his son John Quincy Adams, and George H. W. Bush and George W. Bush) represent a sort of security, and harken back to ancient kingdoms in which a family could be counted upon to rule for generations.

Bloodlines formally ensure and dictate the continuation of royal families. You cannot aspire to be the queen or king of the United Kingdom and the other Commonwealth countries unless you are born or marry into the right family. Until recently, even if you had the requisite bloodlines, you could not be first in line to the throne if you were a woman who had a brother — regardless of your age or competence, or his.

The fixation on blood — be it in the realm of political dynasties or musical stardom — detracts from a level-headed analysis of the quality of a person's leadership or art. By assuming that the traits of leadership or artistry are in the blood, we assume that the people so venerated have a right to lead (or sing), that we should pay attention to them, that they "come by it honestly," that their extraordinary and individual hard work counts for less than their placement inside a family, and that they deserve to be elevated to a godlike status.

And what about the siblings and children who do not climb onto the same podium as their celebrated relatives? Have they disappointed us? Is there something deficient in their characters? What if you attempt to pursue the passion that made your parent or sibling famous, but you turn out to be not very good? There can be the sense that you have somehow violated or disrespected the dictates of your own blood. You have shamed yourself and your family, or so it is felt, and it would have been better if you had not undertaken the endeavour at all. On the other hand, one might say that an untalented child or sibling (or one with a different talent) might have been betrayed by his or her own blood, in that the family magic skipped over that person.

In my opinion, it is smug and self-satisfying to declare oneself special because of family blood. One is no more special because of the blood in the family than one is special by dint of the accident of one's country of birth. Does being a Canadian, or being a member of the Hill family, make me more special — more deserving of privilege, more entitled to comfort, more valuable as a human being, than any person in any country? Of course not. The flip side of egregious pride in one's family blood or luck of citizenship is the sense that others are less human, less valuable, less deserving than you.

Is a person entitled to lead a country because of his or her bloodline? No. The bloodline is a figment of our imagination. A president, prime minister, or dictator's blood does not recirculate in the veins of his or her daughter or son. It's time to move beyond our blood-based obsession with dynasties in politics, and genius in art. Roll over Hippocrates, and tell Galen the news.

ONE NIGHT, WHILE PREPARING to go to bed, my stepdaughter Beatrice Freedman — about six years old at the time — was discussing her own identity with her mom (my wife, Miranda). I was not in the room, but was told that the conversation first touched upon whether Beatrice was Jewish, given that her father was. This is a tricky issue, because Jewish ancestry is traditionally determined by matrilineal descent. Still, Miranda reiterated that Beatrice was related, through her family, to Jewish people. After pondering this for a moment, Beatrice — who had been in my life for about three years at this point — said: "And I'm a little bit black, right?" Miranda asked what Beatrice meant to say. Beatrice replied, "Well, Larry is black and I'm his stepdaughter, so that makes me a little bit black too."

"No, that's not how it works," Miranda replied. She went on to explain that a person's biological parents generally determine black identity. In other words, along lines of family blood.

Eight years have since passed, and in my family we have all had occasion to chuckle about that conversation. But it was more than merely funny. I find it touching to think of Beatrice identifying with people she loved — her father, and me — but running into social rules and barriers every step of the way. Who is to say that Beatrice could not be black or Jewish, if she wanted, and who can argue that others have not done so previously? Is one's identity an absolute function of one's family blood? Can one not create an identity for oneself? Given that identity arises largely from social (as opposed to biological) functions, who is to say that one's identity cannot change — temporarily or permanently? Who is in charge of a person's identity?

We have witnessed many moments in history in which identity gatekeepers held firmly to their posts. During the Holocaust, it was extraordinarily difficult to declare yourself an Aryan, if you were a Jew. Similarly, during slavery in Canada, the United States, and elsewhere, any person of African heritage would have had a huge uphill climb to successfully acquire a white identity.

Even today, some people attempt to control passage through the identity gates. "You're not really black." "You're not completely white." "Where are you from? Yes, but where are you *really* from?" "Isn't your father Jewish?" "Are you a full-bloodied Indian, or mixed?" Many people in various countries have considerable experience answering these questions. In earlier times, being the target of some of these questions might have suggested that your life was in danger. Today, the questions mostly irritate, like flies that refuse to vacate a tent.

I would conclude that identity comes down to a delicate and sometimes ongoing negotiation between the individual in question and those who know him or her. In my own home, my view is that it was right to disabuse my stepdaughter Beatrice of the notion that she might be "a little bit black" given her relation to me. It seemed to me rather obvious that it could set up a white girl for nothing but confusion and trouble if we encouraged her to think that she was black, or partly black. And it seemed only right that she should understand how people, generally, see black identity as deriving from biological family ancestry.

On the other hand, despite the traditional rules of matrilineal descent when it comes to Jewish identity, some people with Jewish fathers (but not Jewish mothers) do indeed identify as Jewish. And others who have no Jewish people in their family convert to Judaism — for a variety of personal reasons, including marriage — and therefore embrace very vigorously their new-found identity. I find the malleability around Jewish identity to be fascinating and inspiring, because it suggests that people can transcend rules of blood descent, when it comes to acquiring or switching their identity.

The late Canadian poet and professor Elizabeth Brewster, one of the founders of the literary journal the *Fiddlehead* and member of the Order of Canada, converted to Judaism. The African-American entertainer Sammy Davis Jr. was born to a Catholic mother and a Protestant father, but converted to Judaism in mid-life after losing his eye in a serious car accident, and never lived down having described himself as a "one-eyed Negro Jew." In 2010, House of Anansi Press published the Canadian writer Alison Pick's novel *Far to Go*, about efforts to transport Jewish children away from danger during the Holocaust. Pick's paternal grandparents had narrowly avoided the Holocaust and, after they settled in Canada in 1941, decided to hide their Jewish identity and not to inform their son (Alison Pick's father) that he was Jewish. Pick's father put the clues together but raised his daughter in a secular family. In adulthood, Alison Pick discovered the family secret and converted to Judaism.

We are led to believe that racial identity is supposed to be clear-cut. But what is simple about the absurd task of defining racial identity? The very idea of race is so contested that some writers — academics in particular — always place the word in quotation marks, as if to say: "Others use this word but I am keeping it between quotes to contest its legitimacy."

Sometimes we think we know all about our family blood, only to be surprised by an inconvenient truth. Catherine Slaney of Brampton, Ontario, grew up believing she was white, until one day she came across a newspaper article — which happened to be written by my father — referring to her great-great-grandfather. She had known that his name was Anderson Ruffin Abbott. She had known that he had been a physician. But she had not heard, until the appearance of the article, that Abbott was black. His descendants had passed so thoroughly for white that it came as a shock to Slaney to learn about a branch of the family that had been kept from her. She embarked on a mid-life quest to identify and meet her black relatives, which formed the basis of her book *Family Secrets: Crossing the Colour Line*.

Slaney's adulthood discovery of her own complex family history hints at the ways that we have come to imagine the very concept of family, and at our eagerness to define it along lines of blood. Much remains to be said about "passing," or slipping imperceptibly from one race into another, and I will return to this issue in chapter five, but it is worth noting that the ways we speak about race and family are unreasonably narrow. For the first many years of her life, Slaney believed that her ancestors were white. And then she learned a fact to the contrary. So how did that discovery change her? What was she, truly? She must have felt that she could walk through a brand new door. Wanting to recognize her family history, but not wanting to be too bold about assuming a brand new identity, she began to refer to herself as a white person with black ancestry.

Step-relationships and adoption also force us to rethink traditional, comfortable notions of blood and belonging in the family. Beatrice and Eve came into my life as "stepdaughters" about eleven years ago, when they were three and six years old. Many parts of their personalities were formed long before they came to know me. As we came to live together, I had to learn to build loving relationships on the fly. We had to cultivate our own connections, even though we had no shared biology. Because their biological parents divorced, and because I'm their stepfather and have a different last name, I had to obtain notarized permissions to take Eve on a trip to Germany, and Beatrice on a trip to Norway. These days, to prevent abductions, when one parent travels alone with a child across a border, he or she generally requires signed permission from the other parent. So I, like others, needed formal paperwork to establish my relationship to my stepdaughters, and my right to travel internationally with them. Obtaining the paperwork, and undertaking the trips helped to cement our very real family connections.

When you declare that you are a child's mother or father, it is assumed that you are saying you share blood. You are thought to share a biology. You helped create the child. And when you announce to the world that you are a child's stepmother or stepfather, some people will assume that you do not count, and that your relationship is not fully authentic. You may have to formally establish your relationship, if you are travelling with a stepchild across an international border. And you may encounter subtle but pernicious rejection when you are picking up a stepchild at a daycare or cheering him or her on in a hockey arena or on a soccer pitch. It may hurt, and it may be annoying or insulting, precisely because you resent another person's value judgement on the quality of your family relationships.

One hardly needs a university degree to observe that whether a relationship is terrific or toxic depends not on biology but on the type of family that you have created. A year or so after my wife, Miranda, and I blended families, I remember being struck to hear Eve stating publicly and proudly that my (biological) son, Andrew, was her brother. She did not call him a stepbrother, because she believed that they were no less siblings than if they had the same biological parents.

I have been much more cautious about dropping the use of the term *stepdaughter*. Beatrice and Eve love and are loved by their father, and when the girls were young I never wanted to give the impression that I was trying to supplant him. They always called me "Larry," instead of "Dad." Over time, however, I have resorted less and less to calling them my stepdaughters, and more and more to just introducing or describing them as my daughters. The precise nature of our relationship, and whether we are "blood relatives," is nobody's business. All people need to know is that we are family.

WHETHER YOU ADOPT A CHILD, blend families with a new partner, or develop any other alternative family structure, you have the ability to say to the world that you believe in the beauty and power of relationships that exist beyond the ties of blood. Blood isn't everything. It doesn't have to be anything. Some people related by blood have such toxic and negative relationships that they would be far better off apart. Most of us can think of a child whom we hope will never have to see a certain parent again. Other people forge deep and enduring relationships with friends they have chosen to declare as family. They know, all too well, that family need have nothing to do with blood, and can have everything to do with choice.

Transracial adoption — domestic or international — complicates this conversation. Many have decried the adoption of Aboriginal or black children by white families in North America, for example, and even likened it to forms of cultural genocide. In Canada, the adoption of thousands of First Nations children into white families has been compared to the impact of residential schools, in terms of the destruction of First Nations culture and identity. In Manitoba, for example, thousands of Aboriginal children were removed from their families and placed in foster care or adopted homes — often in the United States or in eastern Canada — from the 1960s to the 1980s. The problem — dubbed the "Sixties Scoop" — became so severe that in the 1980s, the Manitoba government appointed Justice Edwin Kimelman to oversee a review of Indian and Métis adoptions and placements. Kimelman, a family court judge, declared that people approaching child welfare agencies for help found "their families torn asunder and siblings separated." Child welfare authorities had not intended to perpetuate an injustice, he said, but the forcible removal of thousands of children from their families diminished the lives of individuals and dealt a serious blow to First Nations culture. Kimelman described the placement of thousands of Aboriginal children in white homes outside Manitoba as systemic, routine cultural genocide, and his report in 1985 attempted to stop the trend.

In the United States, the National Association of Black Social Workers (NABSW) has openly favoured in-race adoption and opposed the adoption of black children by white families. In 1972, for example, the NABSW issued a position paper saying: "Black children belong physically and psychologically and culturally in black families where they receive the total sense of themselves and develop a sound projection of their future. Only a black family can transmit the emotional and sensitive subtleties of perceptions and reactions essential for a black child's survival in a racist society." More recently, a public report in 2008 by the Evan B. Donaldson Adoption Institute in the United States echoed some of the same concerns but took a more cautious position, arguing that race should be "one factor, but not the sole factor" in selecting a foster or adoptive parent for a child. "In order for children of color to be placed with families who can meet their long-term needs, consideration must be given to needs arising from racial/ethnic differences," the institute's report said. "Consequently, when workers choose permanent families for children, and when they seek to prepare and support them in addressing the children's needs, race must be one consideration — such as promoting the connection of the child to adults and children from their own racial/ethnic group, developing a positive racial/ethnic identity, and learning to deal with discrimination they may experience."

Clearly, what Edwin Kimelman and the NABSW have in common is a concern that the systematic removal of large numbers of children from a group that is already marginalized is damaging to individual children and destructive to their culture. Even the Donaldson report argued that adoption agencies should recruit families who represent the racial and ethnic identities of children in foster care.

It is a complex issue. On one hand, nobody wants to

see a culture eroded or an individual damaged by being placed in a family where he or she will not be understood or respected. On the other hand, children in need of adoption are often in vulnerable, critical conditions. Their needs must be assessed individually, and according to their best interests.

Just as children need adoptive families, wouldbe adoptive parents need and want children to love. Canadians, Americans, and others in developed countries continue to adopt many children internationally, and transracially.

Many people and organizations openly accept transracial adoption. The Canadian Paediatric Society (CPS) notes that transracial adoptive families need to help children develop the skills they need to be emotionally healthy (an important approach for all parents). In a position paper, the CPS states that there is wide consensus that transracial adoptees, whether domestic or intercountry, need to create an identity that accepts their own physical appearance, their birth heritage, and their heritage of upbringing. And it acknowledges that transracial adoptees and their families do experience racist and stereotypical remarks and are likely to face "racial teasing." Nonetheless, the CPS takes a more positive position on the issue, declaring that transracial adoption affords families an opportunity to experience another culture and to celebrate diversity.

Adoptions cannot be sanctioned if they involve cultural genocide, the theft or sale of children in the wake of social upheavals or natural disasters, or the removal of children against the will of their families. Insofar as international adoptions are concerned, the UN Convention on the Rights of the Child states that inter-country adoption may be considered if it is not possible to place the child in foster care or adoption or to have the child cared for suitably in the country of origin; if the placement does not involve improper financial gain; and if the adoption has the consent of the parents, guardians, or others caring for the child.

If there is respect for the fundamental rights and desires of the child and those who have brought him or her into the world, and if we are not destroying the culture of a country or of a group of people, I would say that adoption — be it international, domestic, transra-cial, or in-race — does offer a unique opportunity. Love is never guaranteed in any family relationship, but adop-tion offers the potential to someone who has not received what many people take for granted — belonging, security, encouragement, and love in a family. It unites people who want to create family relationships, not because of birth relationships or shared DNA, but because of some-thing that runs much deeper than blood: the need for togetherness and bonding in daily life. And it allows us to transcend traditional notions of blood when it comes to family.

There are many ways to parent successfully. And many ways to give a child security, self-esteem, and the means to function as an independent adult. Many biological parents do this beautifully. And so do many step-parents, same-gender parents, and adoptive parents — including those adopting across racial lines. Alternative arrangements give parents and children alike the potential to assert that, when it comes to developing supportive family units, love will trump blood every time. The cliché has it that blood is thicker than water. But it's a ridiculous saying. Who has water in their veins, anyway? Speaking to my own stepdaughter or to my imaginary adoptive child, I would have to say that your blood is no different than mine.

LIKE ADOPTION AND STEP-PARENTING, other cultural traditions and rituals offer the potential to expand the way we imagine family ties. After graduating from university in Canada, my eldest daughter, Geneviève, moved to Ghana to work for a year or so with the World University Service of Canada. When she returned home for a vacation, she noted that in Ghana she would habitually use the term *auntie* in referring to or introducing any woman who was older than her, and *sister* to describe any female friend her age. When I asked if she found it strange to refer to her own boss as "auntie," Geneviève replied that she enjoyed living in a society that did not necessarily share North American obsessions with narrow kinship structures. In Canada and the United States, she argued, people are constantly expected to define the precise biological or blood tie that they share with each other. In Ghana, she argued, the public definition of precise family units counts for much less, and people constantly use the language of family — "mother," "auntie," or "daughter" — to refer to each other, regardless of biological ties.

Just as Ghanaians seem to be telling us that family-like ties do not derive strictly from formal kinship, people of African heritage in the Americas have been calling each other "brother" and "sister" for hundreds of years. When I was a child, I was astounded to see that my father would greet total strangers in the street, if they happened to be black too. In our Toronto suburb in the 1960s, we rarely spotted other black families. "Look Larry," he said to me once, "a black bus driver!" He was showing me, in his own way, that he felt a sense of kinship with them, because they were all part of the African Diaspora.

People can feel the ties of family through a nation or a diaspora, but we also have a long history of forging new familial ties out of thin air. For example, around the world, the notion of blood brothers — unrelated men who create the ties of family by intermingling blood and swearing loyalty to each other — has been yet another way to contemplate the meaning of family ties that reach beyond mere biology.

In one of the most famous medieval legends about blood brothers, the Norwegian warrior Örvar Odd and his Swedish rival Hjalmar convened to do battle. Hjalmar had more boats of war than Örvar Odd, so he set some aside to make it a fair fight. It turned out that they were evenly matched. Instead of continuing on to the death, they ceased their hostilities and formed a pact of blood brotherhood by bleeding jointly onto a patch of earth that was subsequently covered by sod. After creating this pact, Örvar Odd and Hjalmar went on to fight many battles side by side, until Hjalmar died in a duel with another man over the right to marry Princess Ingeborg.

In 1866, the Swedish artist Mårten Eskil Winge painted *Parting from Örvar Odd after the Fight on Samsö*, which shows Örvar Odd bidding farewell to the mortally wounded Hjalmar. Örvar Odd stands in his red cape and clasps the hand of Hjalmar, who lies on the ground with his sword, helmet, and shield strewn beside his body. In the background, a ship glides over the water, Hjalmar's victims lie dead, and birds take flight. Örvar Odd and Hjalmar stare into each other's eyes — brothers to the very end.

In another painting of the very same era by another Swedish artist — August Malmström — Örvar Odd continues to render service to his blood brother, even after Hjalmar's death. The name of the painting — *Örvar Odd Informs Ingeborg about Hjalmar's Death* — says almost everything we need to know. Örvar Odd delivers the body of his blood brother to the woman Hjalmar loved, but Princess Ingeborg cannot sustain the tragedy. She dies of grief in the arms of her father, the Swedish King Yngvi, while Örvar Odd remains strangely immobile, looking on and scratching his beard.

The mythical story of the blood brothers captured the Nordic imagination for centuries and has trickled into global culture so profoundly that the very term *blood brothers* is now part of our daily language. It reflects a global fascination with the idea that brotherhood forged from spilled blood sometimes counts for more than relations resting on a biological foundation.

JUST AS WE HAVE COMPLICATED family relationships, sometimes involving blood and other times moving beyond it, we have a similarly intricate relationship with regard to notions of citizenship. Citizenship is the ultimate expression of group belonging. Citizenship determines where you may live, and very often where you may not live or go without permission. The very notion of citizenship — a defined group of people living in or having access to a given nation, while either keeping others out or limiting their rights as visiting foreigners — divides the earth into arbitrary pieces. On this planet we have massive nations such as Russia and Canada, and tiny countries such as Monaco and Seychelles, but they all have rules about who gets to belong and who does not.

Citizenship also comes with rights, responsibilities, and privileges. Few would dispute that a child born with Canadian, American, British, or French citizenship is likely to enjoy infinitely more privileges and opportunities than, say, a child born with citizenship in Somalia, the Democratic Republic of the Congo, Burundi, Afghanistan, Bangladesh, or Nepal. One current legal case that illustrates the disparity of opportunity revolves around a severely autistic seventeen-year-old girl named

Sabreena Shabdeen, who was born in Canada and has Canadian citizenship, but whose parents were denied refugee status and later moved to the United States. In 2013, Sabreena's parents were at risk of being deported from the United States to Sri Lanka. They expressed the desire to have their daughter moved back to Canada, where she would have better health-care opportunities than in Sri Lanka. At the time of writing this book, the Canadian government had refused to let the parents take their daughter back into Canada — the country of her citizenship — and had recommended that they move back to Sri Lanka. The parents were left with the choice of leaving their daughter alone in Canada (if they could get her to the country), or taking her back to Sri Lanka, where they did not feel she would receive proper care. This would not be an issue if Sabreena's parents had Canadian citizenship. But they don't. So they have no rights in the country.

Traditionally, nations have determined citizenship eligibility by using one of two key principles: *jus soli*, which means that the right to citizenship derives from where you were born, and *jus sanguinis*, which means that citizenship stems from your blood (or ancestry).

In ancient Rome, one acquired citizenship if one was born to Roman citizens, although such matters are never perfectly clear. For one thing, in ancient Rome (and in many countries today, it could be argued), not everybody who was legally established in a nation enjoyed equal rights. Men had more rights than women, for example, and both had more rights than slaves, who were merely another person's property unless they came to be emancipated. But even in Rome the principle of *jus sanguinis* was not absolute. Roman mythology conveys the idea that a person may acquire citizenship by means other than birth. The Greek historian Plutarch made note of this new, expanding concept of citizenship in *Parallel Lives*. Plutarch writes about Romulus and Remus, the mythological founders of Rome, who were said to have been suckled by a she-wolf. (Their fraternal bonding under the ribs of a wolf did not prevent Romulus from later murdering his brother with a stone.)

In *Parallel Lives*, Plutarch wrote that after the Romans abducted and raped the women of the Sabines, Romulus slew Acron, king of the Caeninenses, and then routed the king's army and took his city. And then, wrote Plutarch: "To the captured citizens, however, he [Romulus] did no harm beyond ordering them to tear down their dwellings and accompany him to Rome, where, he promised them, they should be citizens on equal terms with the rest." What interests me about Plutarch's observation is that it shows us that thousands of years ago, nations were expanding notions of what might comprise citizenship, or how a person might become a citizen. Maybe this caught Plutarch's interest because he, a Greek, became a naturalized citizen of Rome.

In modern history, scholars such as the French historian and political scientist Patrick Weil have noted that the principle of birthplace (*jus soli*) determined citizenship in eighteenth-century France and England. However, this changed after the French Revolution. By 1804, the new Civil Code offered citizenship at birth to any child of a French father, living in France or abroad. Other countries, such as Austria, Spain, the Netherlands, Norway, and Sweden, to name just a few, followed suit. But the British principle of *jus soli* took root in Canada, the United States, Australia, South Africa, and other jurisdictions.

In the United States, the Fourteenth Amendment of the Constitution states that "all persons born or naturalized in the United States, and subject to the jurisdiction thereof, are citizens of the United States and of the State wherein they reside." This amendment was adopted in 1868 to ensure that African-Americans would be recognized as U.S. citizens, and to reverse the effects of the 1857 "Dred Scott" case, in which the Supreme Court had ruled that African-Americans could not be considered citizens. This principle of the amendment was tested but vindicated in the case of *United States v. Wong Kim Ark*, which the U.S. Supreme Court ruled on in 1898. This case raised the argument that a certain citizen might not be a full citizen, with the right to return home after travels abroad, simply because of his blood or ancestry.

Wong Kim Ark — who had been born to Chinese parents in San Francisco around 1871 — left the United States and found himself denied re-entry later. Why? At the time, in the face of mounting racism against Chinese immigrants, the federal government was enacting legislation aimed at preventing Chinese people from immigrating to the U.S. The Chinese Exclusion Act of 1882 began with the following lines: "In the opinion of the Government of the United States, the coming of Chinese laborers to this country endangers the good order of certain localities within the territory..." In a split decision, the U.S. Supreme Court ruled in favour of Wong Kim Ark, interpreting the Constitution to mean that almost all people born in the United States would acquire citizenship at birth. In a telling dissenting judgement, the chief justice and an associate justice of the Supreme Court disagreed, arguing that relying too firmly on the principle of *jus soli* would create an untenable situation by which "the children of foreigners, happening to be born to them while passing through the country, whether of royal parentage or not, or whether of the Mongolian, Malay or other race, were eligible to the presidency, while children of our citizens, born abroad, were not."

They need not have worried. *Jus sanguinis*, or citizenship based on blood, also enjoys currency in the United States. A child born abroad of American parents, for example, will enjoy American citizenship — however, he or she is not allowed to become president. To ascend to the country's highest office, you also have to be *born* in the country. (This is not the case in Canada, where four prime ministers — most recently, John Turner — have been born outside Canada.)

The significance of birthplace became the obsession

of the "birther movement" — a group of Americans who vigorously advanced the theory that Barack Obama had been born outside the United States and was therefore ineligible to be president. Business magnate Donald Trump was among those leading the charge. In 2011, Obama released his long-form birth certificate, showing that he had been born, as he had stated all along, in Hawaii. Many African-American and other observers believed that the birther movement sprang from a deep-seated racism — the suggestion that Obama couldn't possibly be a true American or a true president, because he was black.

"Born in the U.S.A." is the title of a 1984 hit song by the singer-songwriter Bruce Springsteen. The hard-driving vocal track, with its repetitive use of the four-word title, sounds like a chest-thumping national anthem, but the lyrics actually lament the dim prospects of American workers who are alienated and struggling after the Vietnam War. For the narrator in Springsteen's song, to be "born in the U.S.A." seems as much a liability as an asset.

In Canada too, citizenship is obtained at birth if one is born in Canada (the principle of *jus soli*), but if one is born abroad of a Canadian parent, one also acquires citizenship (*jus sanguinis*). However, the matter is complicated. In 2009, for example, the Canadian government moved to limit the extent to which *jus sanguinis* applied: a child born outside Canada of Canadian parents will no longer acquire Canadian citizenship if his or her parent acquired Canadian citizenship in the same way (being born outside the country, of Canadian parents). This is an example of the Canadian government moving to limit the application of *jus sanguinis* in order to control the number of people living outside Canada who are to be eligible for citizenship.

Audrey Macklin, professor and chair in human rights law at the University of Toronto Faculty of Law, mentions that if Canadian citizenship rules are meant to include those with a meaningful connection to Canada and exclude those who lack it, then birthright citizenship is clearly an imperfect predictor. "If am born in Canada (*jus soli*) and my parents move to another country when I'm six months old and I never return, I remain a Canadian citizen," Macklin wrote to me by email. "If I give birth to a child outside of Canada, that child will also be Canadian. But if I am born in the U.S. to Canadian graduate student parents, return to Canada when I'm a year old, grow up in Canada, then go study in the U.S. myself and have a baby there, my child will not be a Canadian."

Israel's law of return allows people of Jewish ancestry to move to the country and acquire citizenship, although the same law has drawn criticism for excluding Palestinians. The German *Aussiedler* rule used to allow ethnic Germans in other countries to come to Germany and claim citizenship, even though descendants of Turkish migrants who were born in Germany were not granted citizenship at birth. Many countries today offer a more liberalized application of *jus sanguinis* than

Canada. For example, you may claim Italian citizenship — regardless of whether you speak Italian — if one distant ancestor of yours was a citizen of the country. Many European Union member states allow individuals to claim citizenship if one grandparent was a citizen — a rule which Canadians have often used to acquire an EU passport. Canada's own move to limit the scope of who may be considered a citizen strikes me as a mean-spirited attempt to restrict access to the coveted key of citizenship. It seems to me that you should be able to belong in a country, and to enjoy the rights that go along with it, if you can establish a legitimate connection to that place — by ancestry, by place of birth, by dint of where you have lived for much or your life, or through a combination of those factors.

JUST AS THE UNITED STATES government barred Chinese people from entering the United States and tried to prevent Wong Kim Ark from re-entering the country of his birth, exclusionary policies operated in Canada too. Perhaps most famously, the Canadian government worked assiduously to deport "back" to Japan thousands of citizens (born in Canada, it must be emphasized) of Japanese descent, in response to the Japanese attack on Pearl Harbor and the subsequent war with Japan during World War II.

Although I am speaking of deportation here, because it is the most striking example of citizenship being revoked, it is important to note that regardless of whether they were deported or allowed to stay in Canada after the war, Japanese-Canadians lost the rights of citizenship during the war, because of their ancestry. It didn't matter that many of them had been born in Canada or had become naturalized Canadians. In the time of war, Canadian authorities swept away their citizenship and focused on their race. Their ancestors came from Japan. This, supposedly, meant that they could not be trusted in a time of war with Japan. They didn't just lose citizenship; under Canada's War Measures Act, they were declared enemy aliens. The government confiscated their land and buildings, forced them out of coastal British Columbia, and placed more than twenty thousand of them in internment camps.

The internment and mistreatment of Japanese-Canadians during World War II (which was paralleled in the United States) demonstrates one way that national governments have pointed to blood as a means to justify the mistreatment of their citizens. Who is defined as a slave? Who is black or Aboriginal? Who is detained and searched in the wake of the 9/11 attacks on New York City and the Pentagon? Who is respected in times of war? When crises erupt, blood trumps citizenship.

As Ann Gomer Sunahara notes in her book, *The Politics of Racism*, in 1944 the annual convention of the B.C. Canadian Legion and Vancouver's mayor, J. W. Cornett, demanded that the "Japanese and their children be shipped to Japan after the war and never be allowed to return here." Vancouver Centre MP and federal minister of pensions and health Ian Mackenzie applauded this statement. Despite the fact that both the Supreme Court of Canada and the Privy Council upheld the government's right to deport Japanese-Canadians, Ottawa never managed to deport as many as some had hoped. Sunahara says that 3,965 Japanese-Canadians (51 percent of whom were Canadian-born) were shipped to Japan in 1946. Ostensibly, she says, they departed voluntarily, but in fact most left because the Canadian government had given them no choice.

As the internment and deportation of Japanese-Canadians suggests, citizenship in Canada too is a fickle thing, and in some cases has been revoked as a result of blood. In 1988, the Canadian government finally and formally apologized for interning and deporting Japanese-Canadians during the war. To make amends (mostly symbolic) to individuals wronged during the war, the government set up a modest compensation system for individuals and for the Japanese-Canadian community.

Two recent deportation cases show that the Canadian government is prepared to use whatever principles it can — citizenship as derived by blood or by place of birth — to rid itself of people it considers to be undesirable, even if those same people have a more real and tangible attachment to Canada than to any other country.

In 2012, Canada deported Jama Warsame and Saeed Jama to Somalia. Both were young men and long-time residents, but not citizens, of Canada. Both had been convicted of crimes: Warsame for robbery and drug crimes, and Jama for possessing crack cocaine for the purposes of trafficking. In a 2012 opinion piece for the *Ottawa Citizen*, Audrey Macklin noted: "Both men were born in Saudi Arabia to Somali refugees, arrived in Canada as children, have never set foot in Somalia, and can't speak a Somali language. Saudi Arabia does not grant citizenship on the basis of birth on Saudi soil, so neither are Saudi citizens. They are Somali citizens because they were born to Somali citizens."

The two men had grown up and were educated in Canada, resided in the country but never received Canadian citizenship. Having spent almost all of their lives here, was it right for the Canadian government to treat them and to deport them as Somali citizens? Were their ties to Canada — for better or for worse, and acknowledging the crimes perpetrated — less real than to a country where they have never lived, and don't speak the language?

It is true that the cases of Warsame and Jama do not involve Canadians being stripped of legal citizenship (such as in the case of Japanese-Canadians deported after World War II), but they do show that the government was eager to deport people who had spent most of their lives in Canada and who had real family ties in the country. In both cases, the deportees had been brought to the country as children, but their parents had neglected to obtain Canadian citizenship for them. If Warsame and Jama had been Canadian-born citizens of long-time Canadian ancestors, Canada would be the only country that could claim them. Their undoing — after having been schooled in Canada, and after having learned in Canada how to become criminals — was to be seen as foreigners in the country.

In the case of Jama Warsame, even the UN Human Rights Committee declared that although he did not have Canadian nationality, Canada was still his country. As the committee noted, Warsame "arrived in Canada when he was four years old, his nuclear family lives in Canada, he has no ties to Somalia and has never lived there and has difficulties speaking the language...it is not disputed that (Warsame) has lived almost all his conscious life in Canada, that he received his entire education in Canada and that before coming to Canada he lived in Saudi Arabia and not in Somalia...(Warsame) has established that Canada was his own country...in the light of the strong ties connecting him to Canada, the presence of his family in Canada, the language he speaks, the duration of his stay in the country and the lack of any other ties than at best formal nationality with Somalia."

It didn't matter how forcefully the UN committee argued that deporting Warsame would break Canada's own international commitments as a signatory to the International Covenant on Civil and Political Rights. It didn't matter that in the case of Warsame, the UN committee argued that the man's true country was broader than the mere concept of formal nationality. A person's country, the committee argued, "embraces...an individual who, because of his or her special ties to or claims in relation to a given country, cannot be considered to be a mere alien." In the end, Canada deported Warsame and Jama anyway. They had the wrong blood and had been born in the wrong place.

Citizenship, in my view, should express your true connection to a place. Have you carved out a significant part of your life there? Been schooled there? Established serious family or economic connections? These are just some of the things that bind you to a place. Ascribing citizenship by place of birth, by blood, or by the process of naturalization have all been understandable attempts to shape the notion of who gets to belong. But immigration authorities have ample room to apply these principles in arbitrary or unfair ways in deciding who gets to walk through the door, and who gets to stay.

Just as we still tend to believe that the truest, most authentic family relationships are those based on blood ties, the deportation of Warsame and Jama shows that our nations continue to embrace an age-old notion that citizenship is defined most profoundly by blood. If you are born of the wrong ancestors — perhaps Japanese ancestors during World War II, or Saudi ancestors during North America's post-9/11 war on terror — then you are at risk of being deported in times of war or terror, real or imagined.

I WOULD HAVE A FINE, FAT PIGGY BANK if I were to be paid a dollar for each and every time someone has either told me that I was "half black," or "half black and half white," or has spoken to me about another person having ancestry divided into neatly arithmetical parts, as in "one-quarter English, one-quarter Japanese, and half Tamil."

The arithmetical quantification of race crops up in our daily language, because race is so deeply and subconsciously connected in our minds to blood. I meditated on this numbers approach to blood and race in my 2001 memoir *Black Berry, Sweet Juice: On Being Black and White in Canada*. When you step back and analyze the idea of blood as a metaphor for race, it seems patently ridiculous. Race is an artificial concept. It is an idea that we humans have imposed on one another. Just as one cannot have one knee of a supposed black race and another that is white, or twelve ribs that are Cherokee and another twelve that are Chinese, one can certainly not measure out one's blood in racially distinct one-litre containers.

Race is widely considered to have no grounding in science, but there are various connections between race and scientific investigations. For example, scientists have determined certain DNA markers that tell us if two people are of the same family. Matching people of the same race appears to be one of many factors necessary for complex transfusions affecting material such as bone marrow. Certain diseases are more prevalent among certain racial groups. As well, popular ancestry tests these days can tell a person if he or she has distant ancestors from Africa, and if so, in what regions. However, the fact remains that there is no fundamental biological or genetic difference between people of different races. Genetic diversity is as multifarious and complex among people of the same race as it is across people of different races. Most social scientists these days agree that "race" is a fiction, a sort of social construct, a way that human beings have learned to organize themselves for the purposes of creating social hierarchies, and of justifying injustice. But the way that we have come to understand race consistently and insistently suggests variations in blood parts.

More than two hundred years ago, Western thinkers offered theories of race and biology, and of the supposed superiority of one race — generally called Caucasian — over the others. It would oversimplify matters to finger just one person, but the eighteenth-century German physician and anthropologist Johann Friedrich Blumenbach did not help matters by declaring that humans could be classified into five races: Caucasian (the white race), Mongolian (the yellow race), Ethiopian (the black race), Malay (the brown race), and American (the red race). In his treatise called *On the Natural Variety of Mankind*, Blumenbach had this to say about coining the term *Caucasian Race*: "I have taken the name from Mount Caucasus because it produces the most beautiful race of men. I have not observed a single ugly face in that country in either sex. Nature has lavished upon the women beauties which are not seen elsewhere. I consider it impossible to look at them without loving them."

Caucasian is a strange word indeed, and invokes a confusing geography. These days, people use *Caucasian* almost interchangeably with *white*, but the term comes from people in the Caucasus Mountains, located in Russia but also in Europe, northern India, the Middle East, and North Africa. All sorts of people in the world who don't look white may indeed claim — and have indeed claimed — to be Caucasian in order to gain certain advantages, one of them being the right to become a naturalized American citizen.

Courts have looked dimly on such applications. Bhagat Singh Thind of India lost in the U.S. Supreme Court in 1923, for example, when he asserted that being a high-caste Hindu born in Amritsar qualified him as a Caucasian and thus made him a white person eligible for American citizenship. Justice George Sutherland wrote the decision for the Supreme Court, saying: "We venture to think that the average well-informed white American would learn with some degree of astonishment that the race to which he belongs is made up of such heterogeneous elements." Indeed.

Thinking on a similar level — that if you were from India, you couldn't possibly be Caucasian and you could not be considered a true British subject — played out in Canada too. Consider the *Komagata Maru* steamship, which carried 376 British subjects (including two women and five children) from British India to the port of Vancouver. The ship arrived in Vancouver's Burrard Inlet on May 23, 1914. Some twenty of the passengers were eventually allowed to land in Canada, but the others were refused access, forced to remain on board the ship in deplorable conditions for two months, and eventually required to return to India.

At the time, Canada allowed British subjects to arrive in the country and become citizens, but because the migrants were mostly Sikhs from India, they were deemed not to be British. The B.C. Court of Appeal heard the case in 1914, ruling categorically against the would-be migrants. As James Walker notes in *"Race," Rights and the Law in the Supreme Court of Canada,* lawyer J. Edward Bird argued on behalf of the would-be immigrants from India. Bird said they could not be defined under the category of "Asiatic race" that the Canadian government had permitted itself to exclude from the country on the grounds of being aliens. (In 1910, the Canadian Immigration Act allowed for the exclusion of "immigrants belonging to any race deemed unsuitable to the climate or requirements of Canada.") Bird argued that the migrants were ethnologically Caucasians and racial cousins to the English. He told the judges that the migrants were indeed British subjects, that they should not be categorized as "aliens," and that they deserved to be admitted to Canada, Walker notes.

The B.C. Supreme Court rejected the argument. Writing for the court, Justice Albert E. McPhillips said in the ruling: "It is plain that upon study of the question, the Hindu race, as well as the Asiatic race in general, are, in their conception of life and ideas of society, fundamentally different to the Anglo-Saxon and Celtic races, and European races in general…the laws of this country are unsuited to them, and their ways and ideas may well be a menace to the well-being of the Canadian people."

McPhillips went on to say that an influx of Hindus (as he called them) from India "might annihilate the nation and change its whole potential complexity, introduce Oriental ways as against European ways, eastern civilization for western civilization, and all the dire results that would naturally flow therefrom…their proper place of residence is within the confines of their respective countries in the continent of Asia, not in Canada, where their customs are not in vogue…"

Thus nearly all of the Indian migrants were forced to leave Vancouver Harbour on July 23, exactly two months after they had arrived. When the *Komagata Maru* arrived yet another two months later in Calcutta, police moved in to arrest a number of the men. A riot broke out. The police began shooting, and nineteen passengers were killed.

In 2008, Prime Minister Stephen Harper, the House of Commons, and the British Columbia Legislature issued apologies for the treatment of the people on board the *Komagata Maru*. But the apologies, and the possibility to help nearly four hundred desperate migrants, came ninety years too late. Well into the twentieth century, many British subjects — be they from India, the Caribbean, or elsewhere — were not to be considered true

British subjects and not considered eligible to come to Canada, because they were not white.

The U.S. Supreme Court in its decision about Bhagat Singh Thind, and the B.C. Court of Appeal in its ruling about the Indian migrants aboard the *Komagata Maru*, were influenced — as millions of others have been — by the writings of those who have preached a doctrine of racial distinctiveness and of racial superiority. The courts in both countries declared that if you were from India, you couldn't be deemed Caucasian, British, or American. You could not simply leave your country and waltz into the United States or Canada. You were of the wrong blood.

Working for the Toronto Labour Committee for Human Rights in 1953–54, my mother, Donna Hill, crusaded against federal immigration policy. Canada, as a part of the British Empire, allowed British subjects to come to its lands. But it excluded Indians and blacks from the definition of *British*. Canadian immigration authorities kept a firm grip on who was allowed to enter the country.

In the early twentieth century, the Reverend Samuel Dwight Chown, a Methodist minister and Canada's leading churchman, wrote: "The immigration question is the most vital one in Canada today, as it has to do with the purity of our national life-blood ... It is foolish to dribble away the vitality of our own country in a vain endeavor to assimilate the world's non-adjustable, profligate and indolent social parasites."

Chown's words reflected a wide swath of Canadian twentieth-century thinking, so my mother had her work cut out for her when she went to bat for a twenty-six-year-old Trinidadian man named Harry Narine-Singh, who found himself in a mess of trouble when he showed up in a Toronto army recruiting office in 1954, offering to enlist in the Canadian forces. Narine-Singh, who had been in Canada on a visitor status, was ordered deported after an immigration official by the name of C. Schreiber determined that he was an "East Indian" from "Trinidad, British West Indies." The official later presented Narine-Singh with the explanation that he was not permitted to be in Canada because he was "Asian."

In *"Race," Rights and the Law in the Supreme Court of Canada*, James Walker explains what happened next: "Harry protested that he was not an 'Asian': his family had been in Trinidad for five generations, and he had never even visited Asia; in fact he had never been anywhere except Trinidad and Toronto. Schreiber insisted the Narine-Singhs [Harry and his wife, Mearl] were Asian by 'race.'"

My mother enlisted a lawyer to represent Narine-Singh and his case was fought through the courts, but a year later the Supreme Court of Canada upheld the deportation order. Narine-Singh and his wife were required to leave the country.

I asked my mother, a white American who moved with my black father from Washington, D.C., to Toronto in 1953 and soon became a passionate naturalized

Canadian, what she remembered about Narine-Singh's case. My mother was retired and eighty-five years old at the time of our interview in June 2013, but her mind was clear. She said she didn't remember exactly what was said in court about Narine-Singh, because by the time of the ruling in 1955, she had left her job to have her first baby (my brother, Dan) and begun raising a family. "As a woman at that time, I was not able to have a family and keep my job," she said. My mother had passed along the reins at the Toronto Labour Committee for Human Rights to Sid Blum.

As for what the case said about Canada, my mother got to the heart of the matter: "Canadians always seemed to feel that they were so much better than the Americans because of racism, even though we had it here too. Excluding British subjects who were not white shows that Canadian social attitudes were still really backward at that time, and that many people believed that Canada was basically for whites."

IT WOULD BE NAIVE to conclude that the equation of blood and race is a thing of the distant past, or confined to eighteenth-century European thinkers obsessed by notions of racial superiority. The idea may offend many of us today, but defining people's racial identity along lines of blood — and holding them to it — has been serious business since Europeans began settling in the Americas.

In eighteenth-century Mexico, *criollos* — people of Spanish origin who were born in the New World — were concerned that they were situated one step down the social totem pole from Spanish-born people who had settled in the country. Therefore, they enacted rules regulating and restricting every aspect of the lives of Indians, blacks, and mixed-race people, who were to be positioned several rungs down from them on the ladder of social hierarchy. This caste system, or *sistema de castas*, was designed to ensure that the elite *criollos* were not associated with the tainted blood of miscegenation (the mixing of racial groups through marriage and procreation).

"Mestizos" was a term used to describe Mexican people with mixed Spanish and Indian heritage. Octavio Paz, the late Mexican Nobel Prize–winning poet and essayist, once wrote: "If the criollo, born of Spanish blood, was the victim of ambiguity, the mestizo, born of mixed blood, was doubly so: he was neither criollo nor Indian. Rejected by both groups, the mestizo had no place either in the social structure or in the moral order. In the light of traditional moral systems — the Spanish, based on honor, and the Indian, based on the sacredness of family — the mestizo was the living image of illegitimacy."

As a person of biracial ancestry myself, I'm always a bit wary of the stereotype of the mixed-race person who is forever troubled by the quality of his or her blood. However, there can be no disputing that the most influential classes of people in eighteenth-century Mexico bent over backwards to place peoples of mixed ancestry within a rigid social hierarchy.

Ilona Katzew has written an art history book on

the subject, titled *Casta Painting: Images of Race in Eighteenth-Century Mexico*. Just as in the United States, Canada, Jamaica, Brazil, or any other region where people of many backgrounds found themselves together, it didn't take long for people of one background to end up in bed with those of another background. In Mexico, indigenous peoples, Spaniards, and Africans begin having sex — and babies — as early as the sixteenth century. "This resulted in a large number of racially mixed people known collectively as *castas*," Katzew writes. "By the end of the eighteenth century, approximately one quarter of the total population of Mexico was racially mixed."

The fixation with the *sistema de castas* spilled into the paintings of eighteenth-century Mexico, and this forms the heart of Katzew's book, which contains portraits of the men, women, and children so rigidly defined in Mexico's racially mixed society.

Casta paintings often feature sixteen distinct representations of racial mixing. Sometimes each scene occupies its own canvas or copper plate, and at other times all sixteen are squeezed checkerboard-style into a single frame. Each image shows a man and a woman of two separate races, as well as their child. Each image bears an inscription defining the races of all three subjects in the painting. Juan Rodríguez Juárez, one of the best known *casta* painters, produced bold, lively portraits, each with mother, father, and child. A list of just a few of the inscriptions in his paintings demonstrates the Mexicans' (and the Spaniards') obsession with racial taxonomy. One

is inscribed: "Spaniard and Indian Produce a Mestizo."
Another: "Spaniard and Mulatta Produce a Morisca."
And another: "Mulatto and Mestiza Produce a Mulatta
Return-Backwards," which emphasizes social regression.
(*Mestizo* and *Mestiza* are the Spanish masculine and
feminine nouns indicating people of mixed Spanish and
Indian ancestry. *Morisca* is the feminine noun indicating,
in this case, a person reputed to have one black and three
white grandparents.) There are many names employed
to keep track of all the racial variations: *Albino*, *Coyote*,
Wolf, and so forth. However, there seems to be the possibility to become white again, in Mexico, if a family really
works at it by correcting one step into racial impurity
by another back toward the state of "perfection." This is
shown by another painter, José de Ibarra, who made two
revealing canvases that should be seen back to back. The
first shows "From Mestizo and Spaniard, Castizo." This
castizo child (who, by these arbitrary definitions, would
have one Indian and three Spanish grandparents) looks
confident, happy, and very nearly white. The next intermarriage offers one final possibility of improvement,
showing two finely dressed adults with an aristocratic,
demanding son: "From Castizo and Spaniard, Spaniard."

In case you have trouble figuring out how the arith-
metic equations work between whites and blacks and
their children, a seventeenth-century Spanish Jesuit
missionary by the name of José Gumilla offered a four-
step formula. As Katzew notes, Gumilla wrote:

From European and black, a mulata is born (two fourths each part);

From European and mulata, a cuarterona is born (one fourth mulatto);

From European and cuarterona, an ochavona is born (one eight mulatto);

From European and ochavona, a puchuela is born (entirely white).

Hearing today of the *casta* system and paintings, a reader might declare it absurd and passé. Absurd, yes. Passé, no. We continue to embrace subtler means to define people by dint of their blood purity, or by degrees of mixture. It still influences the way we talk when we describe our neighbour as half black and half white, or our co-worker as one-quarter Irish, one-quarter Japanese, and one-half Nigerian. How quaint. How exotic. How ludicrous. Human identity cannot be arithmetically quantified. You cannot break down blood into discrete racial parts. Each time we try to do it, we reveal the very looniest parts of our hearts and minds. And we open the door to mistreatment and injustice.

People commonly look to slavery in the United States or to apartheid in South Africa for examples of the lengths to which folks will go to establish racial hierarchies. There are more terms for people of African ancestry, and more terms for people of supposedly mixed blood, than could fill a three-hundred-page book. *Mulatto* is one of the most common ones, and it still makes its appearance from time to time these days. It is meant to refer to a person who has one black and one white parent, and is offensive to many because it derives from the Spanish word for the offspring of a horse and a donkey. There is *quadroon* (one-quarter "black blood") and *octoroon* (one-eighth), and from there, the possibilities expand into ludicrous directions. On my shelf at home is one of the most bizarre books I have ever seen, called the *Dictionary of Latin American Racial and Ethnic Terminology*. It is by Thomas M. Stephens, and was published by the University Press of Florida. It has three parts, for terms that are either Spanish-American, Brazilian-Portuguese, or French-American and American French-Creole. It runs to 863 pages. One Creole term is *lèt kayé*, which means "person who is the colour of curdled milk." Another is the Spanish-American term *calpamulato*, which can have various meanings, one of which is "offspring of a mulatto and an Indian; 25 percent white, 50 percent Indian, 25 percent black."

The *Dictionary of Latin American Racial and Ethnic Terminology* shows that as far as people of African descent are concerned, racial identity derives from the idea of blood. In Canada and the United States, the concept of "hypodescent" (also known as the one-drop rule) suggests that a person with any drop of "black blood" would be considered black and so defined for the purposes of slavery, segregation, and other forms of social oppression. Those who would profit from an economy based on the exploitation of slave labour clearly had an interest in defining black people with as wide a net as possible. There is no need to cite a litany of outrageous laws to this effect. One will suffice: in 1934, the State of Tennessee defined the word *Negro* as encompassing "mulattoes, mestizos and their descendants, having any blood of the African race in their veins."

I titled my second novel *Any Known Blood* to refer to a concept that the Swedish sociologist Gunnar Myrdal identified in writing about race in America during World War II. In volume one of *An American Dilemma: The Negro Problem and Modern Democracy*, published in 1944, Myrdal — with his neatly idiosyncratic, foreign phrasing that did such a concise and jolting job of holding up a mirror so that Americans (and Canadians too) could see themselves — had this to say: "Everybody having a known trace of Negro blood in his veins — no matter how far back it was acquired — is classified as a Negro. No amount of white ancestry, except one hundred per cent, will permit entrance to the white race."

Over the course of history in Canada and the United States, people with both black and white ancestry were not excused from the burdens of slavery, segregation, or racial discrimination if they were perceived to have some white blood. In their cases, white blood didn't exist. It didn't matter. It had been polluted. They were judged to be black, and were treated as such, because it was black blood that counted.

The apartheid regime in South Africa built its political

base by separating people along lines of blood superiority (or inferiority, if they were black). South Africa tried to keep those deemed "black," "coloured," and "Indian" separate in every conceivable way from whites. Marriage and sex between blacks and whites was prohibited. There was always the pencil test, for those who wanted to have their race redefined: if a pencil fell straight through your hair to the floor, you could be reclassified from coloured to white. If the same pencil fell out when you shook your hair, you could switch from black to coloured.

This way of thinking was exemplified by the words and deeds of the late American politician Strom Thurmond. An avowed segregationist, Thurmond served as a U.S. senator for forty-eight years. He switched from the Democratic to the Republican Party over his opposition to the U.S. Civil Rights Act. He is famous for having declared, in a speech in 1948: "I wanna tell you, ladies and gentlemen, that there's not enough troops in the army to force the Southern people to break down segregation and admit the Nigra race into our theaters, into our swimming pools, into our homes, and into our churches..."

Why does the absurdity of racial definition become so clear in Thurmond's case? Six months after he died, an African-American named Essie Mae Washington-Williams declared, and proved, that Thurmond fathered her in 1925. At the time, Thurmond was twenty-two, and he slept with and impregnated Carrie Butler, a sixteen-year-old black maid working in his parents' house. Washington-Williams had to attend an historically black college, and Thurmond quietly provided funds for her education. For even a man whose most noted public statement was to insist on racial separation, Thurmond must have known deep in his bones that you cannot keep people apart, and that you bear a responsibility toward the things — and the people — that you help to create.

IT HAS BEEN IN THE economic interests of government agencies to expand the definition of black identity in order to maximize the economic benefits associated with slave labour, but it was not considered such a valuable idea to define all people with Aboriginal identity as "Indians," due to the costs associated with providing services to Aboriginal people or recognizing their land rights.

Virginia's 1924 Racial Integrity Act — one legislative act among many stretching as far back as the 1700s in what would become the United States — referred to blood in defining "colored persons" and "Indians": "Every person in whom there is ascertainable any Negro blood shall be deemed and taken to be a colored person, and every person not a colored person having one-fourth or more of American Indian blood shall be deemed an American Indian."

The same act employed what has famously been described as the "Pocahontas exception," declaring that when the blood was watered down, "Indians" would become white for the purposes of the rules of miscegenation: "It shall hereafter be unlawful for any white person in this State to marry any save a white person, or a person with no other admixture of blood than white and American Indian. For the purpose of this chapter, the term 'white person' shall apply only to such person as has no trace whatever of any blood other than Caucasian; but persons who have one-sixteenth or less of the blood of the American Indian and have no other non-Caucasic blood shall be deemed to be white persons..." I have cited the Pocahontas exception to establish that, over the course of time, authorities have defined racial groups — limiting or expanding the range of people included — to suit their own needs and whims.

Defining Native Americans by means of blood quantum (the quantity of so-called Indian blood in their veins) has by no means disappeared today. Many writers and scholars have weighed in on this matter. One of them, the American Troy Duster, a sociologist at New York University, notes in an article published in 2006 in the *Journal of Law, Medicine and Ethics*, "The U.S. Congress passed the Allotment Act of 1887, denying land rights to those Native Americans who were 'less than half-blood.' The U.S. Government still requires American Indians to produce 'Certificates with Degree of Indian Blood' in order to qualify for a number of entitlements, including being able to have one's art so identified. The Indian Arts and Crafts Act of 1990 made it a crime to identify oneself as a Native American when selling artwork without federal certification authorizing one to make the legitimate claim that one was, indeed, an authentic ('one-quarter blood' even in the 1990s) American Indian."

The problem with equating authenticity to race is that it attempts to quantify a concept that is inherently absurd. To focus on the authenticity of the blood in our veins is to repudiate deeper realities: we construct and negotiate our own identities as acts of social performance from our first to our last days on the planet.

Examples abound in the United States of close adherence to blood quantum in establishing or refuting the identity of American Indians. The Cherokee Freedmen of Oklahoma, for example, have faced an ongoing battle for years to establish that they deserve citizenship and full rights in the Cherokee Nation. Their origin is partially African-American — the Cherokee enslaved their ancestors before the American Civil War. The Cherokee Nation included both Union and Confederate sympathizers. The black slaves of the Cherokee Nation obtained their freedom after the Civil War, and many became members of the Nation. However, in the 1980s, the Cherokee Nation began revoking the citizenship of the Freedmen, unless they could prove descent from ancestors formally listed as "Cherokee By Blood." In 2007, Cherokee Nation voters approved a constitutional amendment that fully stripped the Freedmen of citizenship status.

American scholar Circe Sturm, who is of Choctaw descent, says that — as with most other Indian nations in the United States — you must have a "certificate degree of Indian blood" (CDIB) issued by the U.S. government before you can register as a citizen of the Cherokee Nation. In her book *Blood Politics: Race, Culture, and Identity in the Cherokee Nation of Oklahoma*, Sturm writes: "This small white card, so critical to an individual's legal and political recognition as a Cherokee tribal member, provides some 'essential' information: the individual's name and degree of Indian blood, in fractions according to tribe. For instance, a fairly typical CDIB in Oklahoma might describe someone with multitribal Indian and Euroamerican ancestry in the following manner: seven thirty-seconds Cherokee, two thirty-seconds Kiowa, and two thirty-seconds Choctaw."

Sturm explains that one does not have to have a certain blood quantum to qualify as a Cherokee, but one must establish a link to Cherokee ancestry in a list (called the Dawes Roll) of Indians assembled between 1899 and 1906 by the American government. The Cherokee Nation now includes more than three hundred thousand enrolled citizens, with defined degrees of Cherokee blood ranging from "full blood" to 1/2,048th. The definitions — including the most minute fractions — of Cherokee blood do give credence to the notion of purity. If one has a higher fraction, is one more Cherokee? Is a black person with dark skin more authentically African in origin than a sibling or friend who has African heritage but lighter skin? The very existence of the Cherokee Freedmen attests to the mixing of people of indigenous and African heritage. Does being of mixed heritage make one less African or less Cherokee? I would hope not. The process of mixing cultures should add to your family ancestry, not subtract from it.

In an article in the *Kenyon Review* in 2010, Cherokee scholar Daniel Heath Justice says, "To be Indian in the twenty-first century is to be something other than Indian, as the only 'real' Indians are those locked in museums and nostalgic wet dreams of an imagined and idealized nineteenth-century Wild West." Justice, who is chair of the First Nations Studies Program at the University of British Columbia, asserts in the article that the Cherokee Nation has the right to determine its membership, but that it also has the responsibility to honour "long-acknowledged obligations to those whose ancestors served the Nation in both captivity and freedom."

As for the effort by the Cherokee Nation to exclude the Cherokee Freedmen, some will argue that it is racist; others say that it is right and necessary and at the very least a simple reflection of a long history of formal government policies linking blood and race. I say that the struggle proves that we continue to link blood to race in contemporary society and politics.

Canada too has employed rigid definitions of racial identity. For example, the federal government issued instructions in 1901 to Canadian census takers, with regard to racial identity. The instructions included this explanation: "The races of men will be designated by the use of 'w' for white, 'r' for red, 'b' for black and 'y' for yellow. The whites are, of course, the Caucasian race, the reds are the American Indian, the blacks are the African or negro, and the yellow are the Mongolian (Japanese and Chinese). But only pure whites will be classified as whites; the children begotten of marriages between whites and any one of the other races will be classed as red, black or yellow, as the case may be, irrespective of the degree of colour."

In his 2008 article "From Nation to Population: The Racialisation of 'Métis' in the Canadian Census" in the journal *Nations and Nationalism*, University of Alberta Métis scholar Chris Andersen writes that in the very early twentieth century, the federal government had even more detailed instructions with regard to how to list Native Canadians in the census. Under a category termed Tribal or Racial Origin, Andersen refers to these detailed government instructions to census takers: "[p]ersons of mixed white and red blood — commonly known as 'breeds' — [who] will be described by addition of the initial letters 'f.b.' for French breed, 'e.b.' for English breed, 's.b.' for Scotch breed and 'i.b.' for Irish breed. For example: 'Cree f.b.' denotes that the person is racially a mixture of Cree and French; and 'Chippewa s.b.' denotes that the person is Chippewa and Scotch. Other mixtures of Indians besides the four above specified are rare, and may be described by the letters 'o.b.' for other breed."

Although it is accepted that the Métis people descend from intermarriages between First Nations women and fur traders (of European origins), Andersen asserts that all Aboriginal peoples in Canada — formally listed in our Constitution as Indians, Inuit, and Métis — are of mixed origins. We imagine First Nations people as full-blooded Indians and the Métis people as something approximating half of that, but Andersen argues that the distinctions between Métis and Indian have never been so neat. The derogatory notion of being half-blood or half-breed, so persistently attached to historic Métis identity in Canada, is really not all that different from the reality of Indians, who also have a significant amount of ancestral mixing but whom we tend to imagine as full-blooded or pure — or, at least, more pure than the Métis.

Although the use of the term *half-breed* has been around for as long as Canada has existed, and although the term remained in the Manitoba Act (part of the Canadian Constitution) until 1982, it would be a mistake to think that it has only recently been deemed offensive. Louis Riel, the Métis leader who negotiated Manitoba's entry into the Canadian confederation and who was hanged for treason in 1885 after advocating for Métis rights during the Northwest Rebellion, found the word objectionable.

Canadian lawyer Jean Teillet, a Métis and Riel's great-grandniece, wrote a 2012 annual report called *Métis Law in Canada*, published by her law firm, which quotes Riel on how his people should be described:

> The Métis have as paternal ancestors the former employees of the Hudson's Bay and North-West Companies and as maternal ancestors Indian women belonging to various tribes. The French word Métis is

derived from the Latin participle *mixtus,* which means "mixed"; it expresses well the idea it represents. Quite appropriate also, was the corresponding English term "Half-Breed" in the first generation of blood mixing, but now that European blood and Indian blood are mingled to varying degrees, it is no longer generally applicable. The French word Métis expresses the idea of this mixture in as satisfactory a way as possible and becomes, by that fact, a suitable name for our race.

Riel had more to say on the subject. The intimate, conversational, respectful tone of his words is all the more striking, considering that he wrote his last memoir in a jail cell shortly before his execution. This quote from the memoir comes from Auguste Henri de Trémaudan's book *Hold High Your Heads: History of the Métis Nation in Western Canada,* first published in 1936 and translated by Elizabeth Maquet in 1982:

A little observation in passing without offending anyone. Very polite and amiable people, may sometimes say to a Métis, "You don't look at all like a Métis. You surely can't have much Indian blood. Why, you could pass anywhere for pure White."

The Métis, a trifle disconcerted by the tone of these remarks would like to lay claim to both sides of his origin. But fear of upsetting or totally dispelling these kind assumptions holds him back. While he is hesitating to choose among the different replies that come to

mind, words like these succeed in silencing him completely. "Ah! bah! You have scarcely any Indian blood.
You haven't enough worth mentioning."

Here is how the Métis think privately. It is true
that our Indian origin is humble, but it is indeed just
that we honour our mothers as well as our fathers.
Why should we be so preoccupied with what degree of
mingling we have of European and Indian blood? No
matter how little we have of one the other, do not both
gratitude and filial love require us to make a point of
saying, "We are Métis."

In one key legal case decided by the Supreme Court
of Canada, it became clear how thoroughly we have lost
sight of Louis Riel's reminder that there is no need to
preoccupy ourselves with degrees of blood mixing, and
how profoundly the connection between blood quantum
and Métis identity has lodged itself in the collective consciousness of Canadians.

In 2003, the Supreme Court upheld the constitutional
hunting rights of a Métis father and son, Steven and
Roddy Powley, who had been charged ten years earlier
with shooting a bull moose near Old Goulais Bay Road,
close to Sault Ste. Marie, Ontario. Normally, a person
requires a licence to hunt moose or other animals. The
Powleys had no such licence. They argued that they were
exempt from hunting regulations because section 35 of
Canada's Constitution Act confers Aboriginal and treaty
rights on Indians, Inuit, and Métis peoples. (Canadian

courts had already determined that s. 35 entitles
Aboriginal peoples to fish, log, and hunt for subsistence
purposes, although the Powley case represented the first
time that the Supreme Court of Canada ruled on how
s. 35 applied specifically to Métis people.) The issue at
court was simple: were the Powleys Métis, or were they
not? It was a key point. After all, as the court noted in its
ruling: "The verification of a claimant's membership in
the relevant contemporary community is crucial, since
individuals are only entitled to exercise Métis aboriginal
rights by virtue of their ancestral connection to and current membership in a Métis community."

The case first caught my interest when I was researching *Black Berry, Sweet Juice*, after I heard how forcefully
the Ontario government had raised arguments about
blood quantum in its efforts to see the Powleys convicted.
The Ministry of the Attorney General of Ontario, which
pursued the case for a decade, had argued repeatedly in
lower courts that the Powleys could not be considered
Métis because the father was (in its view) merely 1/64th
Métis and the son an even more scant 1/128th Métis.

In its factum (a set of written arguments) in 1999 to
the Ontario Superior Court of Justice, the Ministry of
the Attorney General argued: "(The) ancestors of the
Powleys married and had children with non-aboriginal
persons of German/English, English and Irish descent.
In particular, Steven Powley's mother (through whom
his aboriginal ancestry is traced) was born near Detroit,
Michigan to parents identified as being Irish and 'white'

or English. Mr. Powley's parents are identified on his birth certificate as being of English and Irish origin. He is of 1/64 aboriginal descent. He married Ms. Brenda Konawalchuk, who is not of aboriginal ancestry. His son, the respondent Roddy Powley, is of 1/128 aboriginal ancestry."

In its forty-two pages of written arguments, the ministry referred no less than three times to the Powleys' blood quantum, concluding: "It is submitted that, at least in a case where there is no evidence of social and cultural ties to an aboriginal community, 1/64 and 1/128 aboriginal descent cannot serve as sufficient 'aboriginal ancestry' to establish Métis identity for purposes of Métis aboriginal rights. That is especially so in a case like this where the aboriginal ancestry of the claimants is traced solely to one individual (i.e., Eustache Lesage, son of a non-aboriginal father and a Métis mother)."

However, the Supreme Court of Canada rejected that argument. "We would not require a minimum 'blood quantum,' but we would require some proof that the claimant's ancestors belonged to the historic Métis community by birth, adoption, or other means," the court ruled. It went on to define just what *Métis* meant, for the purposes of constitutional rights. The term *Métis*, the court ruled, "refers to distinctive peoples who, in addition to their mixed ancestry, developed their own customs, and recognizable group identity separate from their Indian or Inuit and European forebears. A Métis community is a group of Métis with a distinctive collective identity, living together in the same geographical area and sharing a common way of life." To claim Métis status, the court said, a person would have to rely on self-identification, ancestral connection, and acceptance by a rights-bearing Métis community.

In the end, nine years, eleven months, and four days after Steven and Roddy Powley were charged with hunting moose without a licence, the Supreme Court of Canada ruled that the father and son were indeed Métis and acquitted them of the charge.

Although the Supreme Court rejected the concept of "blood quantum," it did emphasize the mixed heritage of the Métis people. I would argue that this continues to set Métis identity artificially apart from that of other peoples. Nobody disputes that the original Métis people of northwest Canada arose as a result of intermarriage between European fur traders and Indian women. But the ongoing emphasis on the mixed heritage of the Métis people suggests that they are distinct in this way, and thus fundamentally different from other people in Canada. The notion is ridiculous. People who are formally defined as "status Indians" in Canada, as well as Aboriginal peoples who lack formal status, are no less mixed in their heritage. Nor are Canadians, generally, any less mixed than the Métis. Métis people *are* distinct from other Canadians. Not because they are more "mixed" than others in the country, but because they developed their own unique blend of people, language, culture, music, arts, lifestyles, and kinship connections.

We continue to nourish an exotic, misplaced, and frequently discriminatory idea about people whose blood is deemed to be mixed, while failing to realize that in the year 2013, a great many of us — Aboriginal, black, white, Asian, and others — are equally mixed.

I would argue that the blood quantum argument, as raised in the lower courts during the Powley case, reflects enduring perceptions — within Aboriginal cultures, and in the broader Canadian community — about the fundamental link between racial identity and blood.

Some sixty-two years before its decision about the Powleys, the Supreme Court of Canada had considered protracted discussions of the racial identity of Indians and Inuit. In the case known as *Re Eskimos*, the court ruled in 1939 that Inuit could be considered Indian for the purposes of deciding that it was the responsibility of the federal government — and not the Québec government — to pay for providing supplies to rescue the Inuit of the Ungava Peninsula in northern Québec from starvation.

In an article commissioned by the federal government for a conference in Halifax in 2001, University of Ottawa law professor Constance Backhouse reviewed the many arguments, pro and con, that had been put to the court about the racial identities of Indians and Inuit. Backhouse observed: "The lawyers for the federal government urged the Supreme Court to draw a legal distinction between 'Eskimos' and 'Indians,' stressing that no one could deny that the 'Eskimo had evolved a distinct civilization and that in physical characteristics, culture, customs, habits and language he forms a group highly differentiated from any of the other aborigines.' The lawyers for Québec, on the other hand, claimed that 'Eskimos' were Indians 'by their blood,' and 'by definition.' 'In a zoological sense,' they argued, 'our eastern Eskimos of the Province of Québec' can reasonably be believed to be Indians 'in bone, flesh and blood.'"

Yet again, identity was reduced to a notion of blood, and the argument appeared to hold sway with the Supreme Court of Canada, which ultimately accepted Québec's arguments, declared that Inuit were Indians, and left the federal government with the tab in a dispute about who should pay to help a starving people.

Canadian legislation such as the Indian Act does not explicitly refer to the term *blood quantum*. Canadians may not use the term as freely as Americans when it comes to defining indigenous identity. However, the principles of blood quantum continue to influence Canadian law and the ways in which some of Canada's 614 First Nations bands consider whether a person's bloodline makes him or her eligible for membership.

Many activists, scholars, and artists, in and out of indigenous communities, have commented on this phenomenon in Canada. Pamela Palmater, a Mi'kmaq lawyer and Ryerson University professor, wrote at length about the Canadian fixation on blood quantum and Aboriginal identity in her 2011 book, *Beyond Blood: Rethinking Indigenous Identity*. Palmater had personal reasons to take on the controversial subject. For years, she and her children were denied formal Indian status — generally required for Indian band membership, and to allow them to assert their rights as First Nations people — because her grandmother had married a non-Indian.

Until the Canadian government revised the Indian Act in 1985, any woman with formal Indian status lost that status if she married a non-Indian. This was not the case for Indian men, whose non-Native spouses (and children) automatically acquired Indian status. When Bill C-31 was passed in 1985 to amend the Indian Act, the Canadian media made much of the suggestion that sexism had been eliminated from the Act. But C-31 merely deferred the problem by one generation. This is because the children of Indian women who "marry out" acquire a second-class Indian status and do not automatically transmit status to their children. For women with this lower level of Indian status to register their children, they must disclose the father's identity and prove his Indian status.

This is how Palmater explained it in *Beyond Blood*:

There are many different ways in which the Indian Act has discriminated against non-status Indians. My grandmother and her direct descendants, from my father, to me, and to my children, all suffer from what is known as "cousins' discrimination," which is a form of gender discrimination. The Indian Act gives lesser status to the descendants of Indian women who "married out"...compared to the descendants of Indian men who married out. We also experience the second-generation cut-off rule, which stipulates that two successive generations of an Indian parenting with a non-Indian will result in no status for their descendants. This kind of discrimination is based on a kind of notional blood quantum allocation...for determining who is really an Indian that is not reflective of actual blood quantum, descent, or culture.

Palmater goes on to note that federal legislation is responsible for creating additional divisions between indigenous peoples by creating two types of status Indians, each named after sections of the Indian Act: 6(1) Indians, who have full status and can transmit their status to their children, and 6(2) Indians, who have half status and who must partner with another registered Indian to transmit Indian status to their children. Palmater writes: "Descent from a single Indian ancestor can be no more than two generations (no less than one-quarter blood quantum) in order to be recognized."

As Palmater says, notions of blood quantum permeate Canadian law even if the term does not appear directly. Essentially, her argument goes, restricting the number of people who can legally be called Indians and pursue their treaty rights has the effect of limiting government costs and steadily reducing the size of Canada's Aboriginal population.

Palmater opens the first chapter of *Beyond Blood* with

an ominous quote from Duncan Campbell Scott, who served as deputy superintendent of Indian and Northern Affairs for Canada's federal government from 1913 to 1932. Scott wrote: "Our objective is to continue until there is not a single Indian in Canada that has not been absorbed into the body politic and there is no Indian question, and no Indian Department." While many Canadians today will cringe at this statement and wave it off as a mentality of yesterday, others will observe that Scott's intent continues to permeate federal legislation to this date.

The National Film Board documentary *Club Native*, by the Mohawk director Tracey Deer, follows four First Nations women who are at risk of being expelled from the Kahnawake reserve (located near Montreal) because they have married non-Mohawk men. This takes place in the wake of a federal government decision in 1985 to allow Indian bands in Canada to determine their own eligibility criteria. Housing and other resources are in demand at Kahnawake, and Indian bands in Canada do not receive federal support for members who are non-status Indians. Some on the reserve feel that "we have to strengthen the bloodline" — these precise words are declared by a band member interviewed for *Club Native*. As a result, the Kahnawake band and many other bands in Canada have adopted restrictive eligibility criteria incorporating the same blood quantum preoccupation that has influenced Canadian laws for centuries.

The rules for how one obtains federally controlled Indian status in Canada, membership in First Nations bands, and other notions of identity — black, Japanese, Chinese, South Asian, white, and others — will all be more respectful and inclusive if we can move away from a fixation on blood, and agree to settle on other ways to define who we are. In 2013, the Federal Court of Canada incorporated this way of thinking into a ruling declaring that Métis and "non-status" Indians are to be considered "Indians" under Section 91(24) of the 1867 Constitution Act. In his decision, Judge Michael L. Phelan wrote: "Degrees of 'blood purity' have generally disappeared as a criterion (for defining a person's Indian status); as it must in a modern setting. Racial or blood purity laws have a discordance in Canada reflective of other places and times when such blood criterion lead to horrific events (Germany 1933–1945 and South Africa's apartheid as examples). These are but two examples of why Canadian law does not emphasize this blood/racial purity concept."

In *Métis Law in Canada*, Jean Teillet argues that the Federal Court ruling correctly defined the Métis as "Indians" for the purposes of Section 91(24) of the Constitution Act, which says that the federal government has exclusive jurisdiction over the affairs of "Indians." However, Teillet says the court wrongly defined Métis people by virtue of their "Indian ancestry" and therefore failed to recognize that the Métis are a separate and distinct Aboriginal people with their own unique identity, language, and culture.

Not surprisingly, the ruling has been appealed to the Federal Court of Appeal. From there, it may go on to be heard by the Supreme Court of Canada. This is certainly not the last time that Canadian courts will weigh in on the matter of Indian, Inuit, and Métis identity. Judge Phelan's reasoning may well be controversial with regard to Métis politics and legal claims, but he certainly appears to have nudged Canada in the right direction by casting aside "degrees of blood purity" as a means of defining human beings.

U.S. PRESIDENT BARACK OBAMA is of black and white ancestry, but can you imagine the ridicule that he would have endured, during any part of his upbringing in the United States, if he had told people that he was white? It would have been a social impossibility. People would have laughed in his face. Black, however, was acceptable. To call himself a black man followed the rule of hypodescent, the one-drop rule, so black it was that Obama had to become.

For nearly four centuries in the United States and Canada, people of African descent have been considered black if they have the very slightest trace of African ancestry. This was accepted by perpetrators of slavery and segregation, and in many ways it became internalized by black people too, who have often been proud and vocal about asserting their identity, even in cases of minimal African ancestry. To deny it, or so the thinking has gone, has been to sell out and to deny a fundamental truth. Moving forward, however, with DNA tests telling blacks that they have some white ancestry (confirming what everybody already knows) and surprising whites with revelations of black ancestry (outing a truth that many already suspect), it may become increasingly impossible — and perhaps it should — to make definitive declarations of race based on blood identity. Genetic testing — which doesn't need blood, by the way, but a mere cheek swab — reveals ancestries either so hidden or so distant that they have become invisible to the human eye. Perhaps in the end genetics will move us beyond blood and race.

If we were not so wedded to the arcane notions of blood, we would be freer to celebrate our various, complex, and divergent identities relating to family and notions of talent and ability, citizenship, and race. We would be more whole, self-accepting people, and less judgemental of others. In this day and age, who among us is not all mixed up?

Arithmetical calculation related to the makeup of blood for the purposes of fixing a person's family, level of talent, nationality, or race is fundamentally demeaning. It boils our humanity down to numbers. It breaks us down into parts, often seizing upon one such part and negating others in order to construct a formal (and artificial) identity. It prevents us from recognizing that it is impossible for us to be half of one thing and half of another, and that it is absurd to suggest that one can be all of one thing (such as black) and none of another (white).

Blood flows efficiently in our arteries and veins, feeding oxygen to our muscles, fighting infection, and regulating body temperature. The magic of blood has the potential to turn toxic, however, when it becomes a metaphor for racial, ethnic, or family identity.

There is no sin in being proud of your heritage and your ancestors. Do you see yourself as African-American? African-Canadian? Jamaican? Korean? Métis? Irish? Sri Lankan? Japanese? Do you remember what your great-grandparents did, and feel that it is important to continue to uphold their values? Do you feel specific, intimate, family-based connections to certain groups of people? It can be a beautiful thing to draw strength and purpose from a sense of ancestral connection.

Interviewed for CBC Radio on June 10, 2013, at the Native Canadian Centre of Toronto, a man named Red Bear, also known as Bernie Robinson, reminded listeners that for many years Aboriginal children were beaten in residential schools for speaking their native languages. Now, said Red Bear, who identified himself as pushing sixty, many Native adults are afraid to relearn their lost indigenous languages because they fear it is too late. But, Red Bear suggested, it is never too late: "Anybody who has Aboriginal blood will pick up that language and it will become part of who they are."

I don't believe that the ability to learn a language is truly located in one's platelets, plasma, and red blood cells. For me, Red Bear's statement evoked the notion of enduring and unbreakable connections to one's heritage and culture. Let's reject the suggestion that blood can be quantified to tell somebody that he can't hunt, or that she can't be allowed to live with her own family and must instead be forcefully removed to a residential school. Let's drop the idea of what you are not allowed to be, or to do, because of who you are, but encourage each other to look for the good in our blood, and in our ancestry. We should let hatred and divisiveness spill from us as if it were bad blood, and search for more genuine and caring ways to imagine human identity and human relations.

FOUR

FROM HUMANS TO COCKROACHES: BLOOD IN THE VEINS OF POWER AND SPECTACLE

WHEN I WAS A PRESCHOOL CHILD, I found it hard to imagine that strangers travelling in cars really, truly had lives of their own. I would be in the back seat of the family Volkswagen Beetle, or perched in the tiny luggage compartment behind the back seat, peering out the windows as my mother or father drove. Who were all those other people in those cars? Did they have places to go to? Their own houses? Their own lives, separate from mine? It seemed inconceivable that such a mass of humanity could exist with no relation to me. I couldn't hear, or touch, or come to know the strangers in those cars, so instead of forcing myself to ponder the ubiquity of mankind, I found it easier to imagine that others who didn't know me didn't really have independent lives at all.

This outlook carried over into my reaction to human suffering. As I began to watch television and movies, I discovered that there were two types of violence on the screen: the kind that I could watch, and the kind that I could not watch. Here was the sort of violence that I could manage, without blinking an eye: Wile E. Coyote gets repeatedly flattened while pursuing his prey, Road Runner, only to rise for yet another indignity a few minutes later. Or, when I was a bit older, the James Bond movie *For Your Eyes Only*, in which an enemy pursuing Our Hero on skis falls into a snow-spraying machine and comes out the other end as bloody snow. Or, even later, action films depicting good guys mowing down dozens of bad guys with gunfire. These scenes did not trouble me, because the victims — perhaps like the strangers I saw travelling in cars when I was very young — remained faceless. They had no humanity. It didn't bother me to see them die.

However, there was another sort of violence that I have never tolerated well. It is when somebody suffers and bleeds up close. To behold their agony while blood flows was, and still is, shocking to me. Before I learned the hard way how viscerally these scenes affected me, I made a fool of myself by fainting once or twice in movie theatres. I went down and out, as if I had been bludgeoned on the head. I fainted watching *Cool Hand Luke*, when the prisoner played by Paul Newman is repeatedly punched in the face by a bigger, stronger inmate. When you follow that scene, you find yourself waiting for the punch that will finally draw blood and end the fight. But the fight does not end with the first bloodshed. It goes on and on. I found myself growing ill as Paul Newman ("Luke") got up and up and up for his repeated beatings. Finally, as he kept getting up, I went down and out.

I had a protected, privileged middle-class childhood with little exposure to real (as opposed to on-screen) abuse and violence in my day-to-day life. But one situation that terrified me, each time it presented itself, was the schoolyard fight. The instant two boys started punching each other, they would be surrounded by a knot of children screaming out, "Fight, fight, fight!" The spectators would stand close together, taut, expectant, demanding results as if they had paid money to watch the spectacle. Once the fisticuffs had begun and chants began to spur the opponents, there was no going back. It was like a volcano. Up and out, the violence had to explode. Those jeering and cheering from the sidelines were not to be denied their thirst to see one person go down — beaten, humiliated, bloodied, and demoted to the very status of the onlooker. There could be only one god — the winner of the fight. The loser would be reclaimed by those watching, and reabsorbed into their midst. His blood flowed right back to where it belonged — alongside theirs, the ordinary mortals'.

I could not understand how children who seemed reasonable in one moment could in the very next demand that blood be spilled. Why would children, as young as seven or eight, call out for bloodshed? Why did they lust to see the victim debased, his humanity flowing out with his blood? What if it had been them? The scenes upset me partly because they made me feel my own vulnerability. Maybe it could be me in that circle. Maybe others would cheer as my own blood spilled.

It became clear to me that crowds gave licence to indecent individual behaviour. The mob had the effect of drawing a mask over the victim. The victim was no longer a person with a sore, broken nose and a mother and father and siblings and pet dog at home. The victim was a faceless person, deserving what he got. The victim was like an enemy assailant in a James Bond movie. Who cared if fifteen bullets tore his body apart? The victim's blood did not matter. It was divorced of humanity. It was blood that didn't count.

Who can forget the nightmare of Lara Logan, the CBS journalist who, while covering the celebration of Hosni Mubarak's resignation in 2011 as the president of Egypt, found herself suddenly attacked and raped in Tahrir Square in Cairo by a large crowd of men? What man, even a brutal sadist and coward, would have dared attack her all by himself, with a huge crowd surrounding him? The massing of men together, in this case, seems to have allowed their evil within to emerge and made it possible for them to carry out the most heinous assault in complete anonymity and with impunity.

In this chapter, I will examine how blood enters the realm of violence, spectacle, and power. Blood inevitably spills when any of these three human phenomena rear their heads. Violence and power need blood. They feed on it, just as cars feed on gasoline. When we want to hurt people, entertain ourselves at their expense, or capitulate to our most base instincts, we lust for blood. People exercising power are not unlike mosquitoes. Female mosquitoes need blood for proteins used to develop their eggs, and they also need the blood of one person to infect another with disease.

THROUGHOUT CIVILIZATION, power has led people to thirst for blood. We have always drawn blood to control, abuse, and annihilate each other. Authorities have repeatedly used the notion of impure blood to whip up fear of individuals or groups, while simultaneously justifying their persecution.

It is widely believed to be the Sumerians — as far back as 4500 BCE, thousands of years before Aristotle was born — who created the legend of Lilitu, the demon who preyed on sleeping men, made off with their blood, and murdered babies and children. Lilitu is considered by many to be the prototype of the Hebrew Lilith (the first wife of Adam), who figures in Jewish demonology and is mentioned as a demon in the Dead Sea Scrolls, which are considered the oldest surviving manuscripts of material appearing in the Old Testament and in other ancient texts. The figure of Lilith has a low profile in the Old Testament. She is not yet a full-fledged witch, but she will evolve into the role later, in the human imagination.

In *A New History of Witchcraft: Sorcerers, Heretics and Pagans*, Jeffrey B. Russell and Brooks Alexander write that the Sumerian Lilitu "was a frigid, barren female spirit with wings and taloned hands and feet; accompanied by owls and lions she swept shrieking through the night, seducing sleeping men or drinking their blood."

The image of the demon Lilitu still lingers today. The contemporary New York visual artist Tara McPherson's painting *Lilitu*, completed in 2010, depicts a red-haired, black-skinned female demon, naked in a thigh-deep pond, holding the bleeding head of a decapitated man, with poisonous drops arching away from her nipples. You are left with the impression that Lilitu seduced the man before she dispatched him. On her website, McPherson writes: "This painting is based on the Sumerian myth of the demon Lilitu. Representing chaos, seduction and ungodliness. She is a sexually charged yet infertile succubus who behaves aggressively toward men and children killing them at her every whim, her breasts are filled with poison not milk. In her every guise, Lilitu has cast an evil spell on humankind."

Blood — or the lack thereof — is a central image in the depiction of the early witch. In the imagination of their persecutors, witches become both irrepressibly sexual bloodsucking fiends who prey on the bodies of sleeping men, *and* old, frigid, bloodless, crabby hags who cause plagues, destroy crops, and bedevil mankind. The witch seeks the power and vitality of human blood because she herself lacks the life-giving substance in her veins.

Fears about witches were fed by the portrayal of women as uncontrolled sexual deviants, both in mythology and in actual historical accounts of ancient Greece and Rome.

In Greek mythology, Dionysus, the god of winemaking and ecstasy, attracted throngs of wild female followers. The maenads, or "raving ones," also known as the bacchantes in Roman mythology, entered into mad, drunken frenzies in which they would rip apart a bull (the symbol of their god) with their bare hands and then eat its flesh and drink its blood. Peter Paul Rubens is well known for his painting *The Bacchanal*, which depicts fauns — the upper halves of their bodies human, and the lower halves goat — cavorting naked and drunk with devils and a black woman. Another painting, *Bacchanalia*, by the twentieth-century Belgian artist Auguste Leveque, shows men and women who are members of a cult of Dionysus. Although known for excesses of orgy and cannibalism, the bacchantes are rendered in a more tender light in the 1887 painting *The Women of Amphissa*, by the British painter Lawrence Alma-Tadema. In this painting, the bacchantes — all young women — find themselves in the central Greek town of Amphissa, where they have gathered to celebrate Bacchus. They are exhausted after a night of revelry and have fallen asleep at the very time when the soldiers of an invading army might ravish them. In the painting, market women gather around the sleeping or sitting bacchantes to offer them protection and food.

Out of the mythological stories of the Greek god Dionysus, or his Roman equivalent Bacchus, and his sexually aroused, debauched followers grows a tradition called the Bacchanalia in which people engage in orgy and wild, drunken celebrations. The Roman historian Livy wrote accounts of orgiastic festivals carried out by those celebrating the Greek god. Russell and Alexander note that the Roman Senate outlawed the Bacchanalia in 186 BCE, but the idea of rabid, rampant sexuality, blood lust, and ritualistic murder began to become associated with the idea of witchcraft. Both the classical Greeks and Romans considered magic and sorcery a menace to society. The writers of the Bible were concerned about witchcraft too. Exodus 22:18 says: "Thou shalt not suffer a witch to live."

Over the centuries, Europeans became increasingly obsessed with the notion of witchcraft, until fears and rumours reached their peak in a four-hundred-year period beginning around 1400. Estimates vary on the number of people accused, tried, tortured, and executed during the heyday of witch trials. In *Witchcraft: A Very Short Introduction*, Malcolm Gaskill says that between 90,000 and 100,000 trials were conducted in Europe, Scandinavia, and America between 1400 and 1800. Russell and Alexander offer higher numbers, suggesting that between 1450 and 1750, some 110,000 people were tortured as alleged witches, and 40,000 to 60,000 were executed. This includes the persecution of witches in America through incidents such as the Salem witch trials of 1692, in which nineteen people were executed and more than one hundred jailed.

Regardless of the exact number of witches tortured or killed, most of those accused and murdered were women, and much of the lore related to the danger of witches sprang from concerns about blood. Certain descriptions and beliefs recurred. Witches were old women. They consorted with the devil, and slept with him too. They put their supernatural powers to maleficent use, causing every manner of trouble, such as hailstorms, lightning, crop failures, the madness of horses, female infertility, and male sterility. They met in assemblies ("Sabbaths" or "sabbats") at midnight.

As early as the middle of the fifteenth century, we have sketches of witches airborne on brooms. For example, the French poet Martin Le Franc's *Le Champion des dames* — in which two people converse about witches and witchcraft — offers a marginal illustration of two women flying through the air. One sits astride a broom, and the other, a simple stick. This dates back to about 1440–42.

In the same era that the Catholic Church demonized and persecuted Jews and Muslims through the Spanish Inquisition, anti-witch inquisitors wrote two documents that influenced European thinking for centuries with regard to women, witchcraft, and blood. The first, written in the 1430s, was penned by an inquisitor of the Duchy of Savoy. The original Latin version was known as *Errores Gazariorum*. It is known in English as *The Errors of the Gazarii* or *The Errors of the Cathars or Waldensians*. As Malcolm Gaskill notes, the treatise describes heretics meeting secretly at night to carry out ghastly behaviours that would later become distinctly associated with witches. *The Errors* tells of heretics misbehaving at "sabbats" and in "synagogues." It alleges that these heretics kill children, bring their bodies to the "sabbat," or assembly, to be eaten, and make lethal powders from their remains. In addition, *The Errors* says, the devil keeps a list — written in blood — of each heretic who joins the pact.

Just a few decades later, the German Heinrich Kramer — an inquisitor for the Catholic Church and a member of the Dominican order — wrote one of the most enduring and influential diatribes about the devilry of witches. Entitled *Malleus Maleficarum* in Latin (*The Hammer of Witches* in English), it warned inquisitors about the havoc wreaked by witches. Women, he said, fell into the practice of witchcraft as a result of debased character and uncontrolled sexual desires.

In *The Hammer of Witches*, Kramer wrote: "... certain witches, against the instinct of human nature, and indeed against the nature of all beasts, with the possible exception of wolves, are in the habit of devouring and eating infant children ... a certain man had missed his child from its cradle, and finding a congress of women in the night-time, swore that he saw them kill his child and drink its blood and devour it ..."

The Hammer of Witches was repeatedly reprinted, and it influenced European thought about witchcraft for the better part of two centuries. Ever fearful of plagues, natural disasters, hunger, and the angry hand of God, humans imagined the devilry of witches to find a convenient scapegoat. The persecution of witches declined in the decades following the Salem witch trials. That year — 1692 — marked the last time that a witch would face execution in America. According to Russell and Alexander, no witches were executed after 1684 in England, 1745 in France, and 1775 in Germany.

Although the trials and persecution of witches diminished after the eighteenth century, the notion of the witch as a bloodsucking, cannibalistic predator lives on in our collective imagination.

In 1812, the Brothers Grimm made famous the story of Hansel and Gretel. The siblings meet a witch who abducts them and begins to try to fatten up Hansel with a view to roasting and eating the boy. She attempts to roast and eat Gretel too, but the quick-witted girl shoves "the Godless witch" into the oven, closes the door, and burns her to death.

In *The Wonderful Wizard of Oz*, written by L. Frank Baum and first published in 1900, we come a little closer to our theme of blood. The Wicked Witch of the West, who is the most formidable enemy faced by Dorothy, the Tin Woodman, the Scarecrow, and the Lion, strikes Dorothy's dog, Toto, with an umbrella. In return, the perky dog bites the witch's leg. Baum writes: "The Witch did not bleed where she was bitten, for she was so wicked that the blood in her had dried up many years before."

The lack of blood in the Wicked Witch of the West indicates her lack of vitality and humanity. Several paragraphs later, Dorothy stumbles upon a method of killing the witch. Angry that the witch has stolen one of her ruby slippers, Dorothy dashes a bucket of water on the aquaphobic witch, which causes her to melt into a shapeless mass and die.

We shouldn't be surprised to see the Wicked Witch of the West succumb so easily after being dashed by water. More than a century earlier, the celebrated Scottish poet Robert Burns plainly noted the inability of witches to handle water. In his 1790 narrative poem "Tam o' Shanter," Burns shows the hard-drinking Tam (or Tom) fleeing on his grey mare Maggie after spying on a group of cavorting, murderous witches. The horse saves Tam's life, but the grey mare pays a price: a witch in hot pursuit is unable to chase Tam across the river but manages to bite off the horse's tail. In a footnote to the poem, Burns writes: "It is a well-known fact that witches, or any evil spirits, have no power to follow a poor wight (or unfortunate soul) any further than the middle of the next running stream." Thank you, Robbie Burns.

The process of dehumanizing people — and women in particular — enabled us to persecute them for centuries, and to convert them into witches. To safeguard society, we spilled their blood even as we accused them of doing the same.

In our minds, those accused of witchcraft suck the blood of others, but have none of their own. They are decrepit and dehydrated, and die when exposed to water, which is the ultimate cleansing, replenishing, life-sustaining fluid. Vilify a woman's bodily fluids, declare that what cleans and hydrates the rest of us is what kills her, and behold, you have created a witch.

ALTHOUGH THE IDEA OF A WITCH as a bloodsucking female demon is a figment of human imagination, there are numerous examples in contemporary society of the violent removal of human blood. Some of the approaches are unethical and illegal, and others are sanctioned and carried out by the state.

During the Holocaust, the Nazis subjected concentration camp victims to every manner of medical experiments. In addition to conducting surgeries without anaesthetic, infecting people with germs and bacteria, exposing them to mustard gas, and injecting them lethally, they also withdrew blood forcibly. Josef Mengele, the infamous doctor of the Auschwitz death camp, transfused the blood of twins and carried out other experiments on their blood. Many of his victims were children, most of whom were killed once the experiments were over.

The forcible and blatantly criminal removal of blood does not end with the Holocaust. In his 2011 book *The Red Market: On the Trail of the World's Organ Brokers, Bone Thieves, Blood Farmers, and Child Traffickers*, investigative journalist Scott Carney writes that the illegal harvesting of body tissue, organs, and blood is more pervasive and profitable now than at any other time in human history. He provides details about a farmer in India who held male captives in makeshift prisons on his property, repeatedly withdrawing their blood, which he sold to doctors, blood banks, and hospitals for fees ranging between $20 and $150 a pint. Carney says the farmer was convicted and jailed after one of his captives escaped and alerted police. When police came to the farm, in the city of Gorakhpur on the border of Nepal, they freed seventeen men. "Most were wasting away and had been confined next to hospital-issued blood-draining equipment," Carney writes. "In their statements the prisoners said that a lab technician bled them at least two times per week. Some said that they had been held captive for two and a half years. The Blood Factory, as it was quickly known in the press, was supplying a sizable percentage of the city's blood supply and may have been the only thing keeping Gorakhpur's hospitals fully stocked."

Carney provides fleeting mention of just one other underground blood farm in India — an operation in Calcutta in the 1990s — but his point is well taken. As long as hospitals are willing to buy blood from unscrupulous dealers, criminals will have an incentive to abduct and detain victims and to withdraw their blood forcibly and for profit. In some places in the world, blood means business.

But forcible blood removal is not always deemed illegal. In North America, it is common for police, or for authorities acting on their behalf, to withdraw blood without consent from people suspected or convicted of criminal offences. In Canada, authorities who obtain a judicial warrant may withdraw blood forcibly from a person suspected of having committed a serious crime such as murder, rape, or arson. In limited cases, without even having a warrant, police may ask a medical professional to forcibly withdraw blood from a person suspected of impaired driving.

In addition, Canadian judges are required to order the taking of blood, buccal or hair samples from people convicted of serious, violent crimes. Since 2000, information from these DNA tests is fed into the National DNA Data Bank used by Canadian police forces. It seems Orwellian to me, but nobody is objecting loudly. In the collective psyche of people who wish to be shielded from crime, it may be an invasive procedure but it serves a greater good. The RCMP website proudly states: "The National DNA Data Bank is a shining example of the increasing importance of science and technology in modern law enforcement... To stay ahead of the criminals, we must make better use of cutting edge science such as forensic DNA."

During his trial in 1997 in Edmonton for first-degree murder, Peter John Brighteyes complained that his constitutional rights had been violated when police took his blood to gather evidence. Brighteyes' lawyer argued in court that the involuntary blood test should be excluded from evidence because the security of the body "is perhaps the most fundamental aspect of security of the person." The Crown argued, successfully, that it needed DNA evidence in Brighteyes' blood sample to prove that he had raped his victim before he killed her. The court allowed the evidence to be used, and sentenced Brighteyes to life imprisonment. Brighteyes, who had a history of violent crime, later hanged himself in prison.

In the United States, forcible blood extraction has extended to hospitals, police stations, and even roadsides, in cases where police suspect people of operating motor vehicles under the influence of drugs or alcohol. Indeed, some American police officers have even been trained in the practice of phlebotomy, or bloodletting, so that they can carry out the procedures quickly. The thinking is that they have to act quickly to gather evidence, since alcohol dissipates quickly in the bloodstream. The blood test is seen as the gold standard, offering more accuracy and more information than the standard Breathalyzer test. In some cases, police go straight for blood.

American courts have stepped in to establish certain limits, warning authorities to respect the Fourth Amendment to the Constitution, which aims to protect citizens from unreasonable searches and seizures. In 2008, for example, a superior court judge in Pima County, Arizona, barred the use of blood-alcohol evidence in the case of a Tucson resident whose blood had been drawn by one cop in the back seat of a police cruiser, while another cop held up a flashlight. "Romantic though it may sound, phlebotomy in the back seat by the dashboard lights is, in this humble trial judge's opinion, unconstitutional," Richard S. Fields wrote in his decision.

In the *Missouri v. McNeely* case in 2013, the U.S. Supreme Court upheld the privacy provisions of the Fourth Amendment to the Constitution and ruled that law authorities may not *routinely* take blood from people suspected of driving impaired. However, the court said, in impaired driving cases, blood may be taken forcibly with a warrant, or when exceptional circumstances justify taking it without a warrant. American courts have also upheld the rights of prisons to withdraw blood forcibly from inmates.

Clearly, the advent of laboratory testing for alcohol, and forensic testing used for criminal investigation, is widening the avenue of state intervention in our bloodstreams. We may not like the intrusions upon our privacy, but lawmakers and courts in North America are clearly lining up on the side of what they consider to be the greater good. They believe that convicting criminals is more important than the privacy of our bloodstreams. You are almost certain to have no control over who gets to withdraw and inspect your blood, if you are reasonably suspected of having committed a violent crime. But driving under the influence of drugs or alcohol could lead to the same result. So watch out, if you are drinking and driving. While a police officer looks on, you may lose more than your driver's licence.

WHEN WE SEEK TO UNDERSTAND why blood has come to be inextricably linked to the manifestation of power and spectacle in society, it is always revealing to look at the Bible. It teems with references to the power and centrality of blood, as an expression of humanity and a vehicle with which to respect and approach God. In the Old Testament, Leviticus contains numerous edicts about blood as it relates to animal sacrifice, food preparation, menstrual blood, and sex. It declares that one must not eat blood, one must remove it from meat that is to be cooked, and a husband must not be intimate with his menstruating wife. Blood, it tells us, is powerful.

In Leviticus 17:11, God tells Moses: "For the life of the flesh is in the blood: and I have given it to you upon the altar to make an atonement for your souls: for it is the blood that maketh an atonement for the soul."

The New Testament emphasizes the necessity of bloodshed to expiate sin: "Indeed, under the law almost everything is purified with blood, and without the shedding of blood there is no forgiveness of sins." (This line comes from Hebrews 9:22, which is distinctive among the New Testament books for its obsession with sacrifice and the temple.)

Even the German philosopher and writer Friedrich Nietzsche, who challenged the very foundations of Christianity and famously declared that "God is dead," seems to have been transported by the transcendental quality of blood. In his novel *Thus Spoke Zarathustra*, Nietzsche wrote: "Of all that is written, I love only what a person hath written with his blood. Write with blood, and thou wilt find that blood is spirit." Nietzsche was not referring to bloodshed explicitly, but telling us that mimicking the rhythm and pulse of blood in the body is what makes writing sound and feel authentic. Words should flow just as naturally and undeniably as the beating heart.

Centuries before *Thus Spoke Zarathustra* first appeared, around 1885, the renowned seventeenth-century Mexican poet Sor Juana Inés de la Cruz applied the dictum literally. For any person who treasures the ascent of the intellectual and creative mind, it is painful to behold the rise and fall of Sor Juana. Her story reflects how powerfully the Catholic Church controlled women who gave their lives over to God. Despite having spent her entire adult life in a convent in Mexico City, the nun had become famous throughout the Spanish-speaking world for her poetry. Today, she is widely considered the greatest Spanish-language poet of her era. However, near the end of her life, in a complicated series of machinations described by her biographer Octavio Paz, male church leaders mounted an attack on Sor Juana's character that eventually overwhelmed her. One year before dying of the plague, Sor Juana renounced her "humane studies" — abandoning her lifelong passion for writing, and relinquishing her personal library of some five thousand books, as well as her musical and scientific instruments — by signing an abjuration with her own blood.

In the renunciation, Sor Juana promised "to follow the road of perfection," by which she means the worship of God to the exclusion of her more earthly passions. She states her belief in Catholicism, says that she is pained exceedingly to have offended God, and reiterates her belief in Mary's immaculate conception. And it ends with these words: "And as a sign of how greatly I wish to spill my blood in defense of these truths, I sign with it." Blood symbolized for Sor Juana the renunciation of intellectual and creative pursuits. And blood bound her to a promise to devote the rest of her life to God, and God only.

The illegitimate daughter of a Spanish captain and a mother who was a *criolla* (of European ancestry but born in the Americas), she was born near Mexico City between 1648 and 1651, and given the name Juana Inés de Asbaje y Ramírez de Santillana. Sor Juana, as she is commonly known, showed astounding intellectual ability from an early age. She was said to have been reading and writing by the age of three, crying out in verse when spanked, and teaching Latin by the age of thirteen. In her famous essay and autobiography *The Answer*, she wrote: "Though I was as greedy for treats as children usually are at that age, I would abstain from eating cheese, because I heard tell that it made people stupid, and the desire to learn was stronger for me than the desire to eat."

Michael Schuessler, a professor in the Department of Humanities at the Universidad Autónoma Metropolitana, wrote about Sor Juana in the essay "The Reply to Sor Philothea," published in 1999 in *Latin-American Literature and Its Times*. "Essentially," Schuessler wrote, "there were three 'career possibilities' for women of seventeenth-century New Spain. The first and most common was that of wife and mother; the second — less common, but far from disrespected — was that of the habit, life as a nun. The third was prostitution..."

Sor Juana chose to be a nun, entering the Convent of San Jerónimo in Mexico City in 1669. As noted by Octavio Paz, her most celebrated biographer, in the book *Sor Juana; or, The Traps of Faith*, "abruptly, she gives up worldly life and enters a convent — yet, far from renouncing the world entirely, she converts her cell into a study filled with books, works of art, and scientific instruments... She writes love poems, verses of songs and dance tunes, profane comedies, sacred poems, an essay in theology, and an autobiographical defense of the right of women to study and to cultivate their minds. She becomes famous, sees her plays performed, her poems published, and her genius applauded in all the Spanish dominions, half the Western World."

In *The Answer*, Sor Juana wrote that in the convent, when her inclination to study was "snuffed out or hindered with every (spiritual) exercise known to religion, it exploded like gunpowder; and in my case the saying 'privation gives rise to appetite' was proven true."

Émile Martel is an award-winning French-Canadian poet from Montreal who has translated Sor Juana's work into French in a book called *Écrits profanes: un choix de textes*. Martel told me that his favourite work by Sor Juana is a sonnet about a woman who sees a portrait of herself and then struggles to criticize, minimize, and deny her own beauty. In an English translation by Electa

Arenal and Amanda Powell, the sonnet carries a rather long but intriguing title: "She endeavours to expose the praises recorded in a portrait of the Poetess by truth, which she calls passion."

In it, Sor Juana speaks of her own portrait as a "cunning trap to ensnare your sense." There may well be beauty amidst the "clever arguments of tone and hue." But, ultimately, admiring attractiveness in a portrait is

> a foolish, erring diligence
> a palsied will to please which, clearly seen,
> is a corpse, is dust, is shadow and is gone.

In forcefully denying her own beauty (Octavio Paz describes her as having been beautiful), Sor Juana imagined her own destruction. She met her downfall, and paid for it with her own books — and blood — after infuriating the men who ruled her world of Catholicism. She had already generated petty jealousies among nuns who watched her become famous throughout the Spanish empire. At least once, they succeeded in preventing her from reading for a period of months. But her undoing — one of the traps of her faith, as Paz puts it — was set in motion when she defended the right of women to study, teach, and write in *The Answer,* an ardent response to a bishop (posing as a nun) who had criticized her for paying too much attention to earthly matters and too little attention to her faith. Sor Juana demolished the bishop's arguments one by one. To justify her own lifelong explorations of literature and science, she cited numerous people of faith — male figures such as Saint Augustine and Saint Jerome, and female figures such as the Queen of Sheba, the Roman goddess Minerva, and the Swedish Queen Christina Alexandra — who had done the same. But her response angered the men who ruled her world, and they saw to it that she was stripped of her earthly pleasures.

Sor Juana died of the plague after treating others who succumbed to the same fate. Octavio Paz notes that at the time of her death in 1695, the disease was commonly treated by means of bloodletting. In addition, he writes, as a member of the convent, Sor Juana would likely have been required to engage in self-flagellation — as did other nuns — in times of plague. "As they wound their way through the corridors and patios of the convent, singing and praying, they scourged and lashed themselves," Paz writes. "A Christian version of the bacchantes and mae-nads: the nuns in the night shadows, half naked, bodies bleeding, singing and wailing."

It is sad to imagine the seventeenth-century nun drawing her own blood to renounce the life of letters. Did she cut a finger on the hand used for writing? Did she nick her arm or leg? Was she alone in her study, reaching for a private, hidden section of skin never seen by others? Sor Juana does not say where or how she cut herself, and does not speak of any pain it might have caused. I imagine she interpreted any such self-inflicted wound as a way of doing penance, prostrating herself before her maker. It seems to dovetail perfectly with Paz's description of nuns flailing themselves and walking through the convent to wail and bleed under the cover of darkness. Some things feel safer in the absence of light. Removing our clothes. Kissing. Making love. And, for the Roman Catholic sisters of seventeenth-century Mexico, bleeding to pay for their sins.

Sor Juana has been the subject of countless essays, dissertations, and books. In 1990, the late Argentinian filmmaker María Luisa Bemberg released a feature film about Sor Juana, entitled *Yo, la peor de todas*. (In English, that means "I, the worst of all the world.") The film's title borrowed from some of the last words that Sor Juana wrote. They came after her renunciation of writing. After she gave up her books. As Paz notes in *The Traps of Faith*, a few months before she died, Sor Juana entered these lines into her convent's Book of Professions: "In this place is to be noted the day, month and year of my death. For the love of God and his Most Holy Mother, I entreat my beloved sisters the nuns, who are here now and who shall be in the future, to commend me to God, for I have been and am the worst among them. Of them I ask forgiveness, for the love of God and his Mother. I, worst of all the world, Juana Inés de la Cruz."

Why did Sor Juana go from being celebrated throughout the Spanish empire to being without books or writing implements in her final years, whispering toward the grave that she was the worst of all the world? Octavio Paz invokes two reasons. The first, he says, is the opposition between the intellectual life and the duties and obligations of the convent life. "The second is the fact that she was a woman," he states. "The latter was the more decisive; if she had been a man, the zealous Princes of the Church would not have persecuted her."

SOR JUANA SIGNED HER LETTER with her own blood to symbolize her sacrifice to God. Bloodshed usually shocks people. To be made socially acceptable, it must be marked by ritual. What might be illegal in one context — say, punching somebody in the eye and causing him to bleed profusely — will be cheered, praised, and remunerated if it unfolds during a heavyweight boxing match. Creating rules about bloodshed lets us quench our thirst for violence without self-castigation or concern that we are giving in to our most base instincts.

I want to examine the rituals governing bloodshed as a way of remembering that we always walk a razor's edge when it comes to what is characterized as civilized or uncivilized behaviour.

Consider the tradition of duelling. A long-time indulgence among aristocrats and the wealthy in Europe and North America, duelling was bound by rules of blood. The duel was a strange thing. If you were a gentleman, you played by the rules — and there were plenty of them. You never challenged another man to a duel for pleasure, but rather, to correct an insult to your honour. If someone called you a fool, slapped your face, or was rumoured to have slept with your wife, then, by George, bring

out the duelling swords. (Or, after the eighteenth century, pistols.) You had to select your duelling weapons,
which were supposed to be equal and lethal. On May 3,
1803, one pair of duelling men fought from gas balloons
floating some nine hundred metres above the Tuileries
Garden in Paris. The duellists fired at each other with
guns. One was reported to have "fired his piece inef-
fectually," but his antagonist proved to be a better shot.
He pierced his adversary's balloon, which crashed down
onto a house below, killing the opponent and his second.
(In the world of duelling, a "second" is the person who
escorts the duellist to and from his confrontation, and
sees to it that the rituals are followed.)

Although the duel in balloons above the Tuileries
ended the lives of two people, many other duels fought
on the ground allowed an antagonist to save his honour
and life by drawing his opponent's blood, or by having
some of his own drawn. At the first sight of blood, opponents were offered the opportunity to step back from
hostilities and consider the matter resolved. Thus the
notion of drawing "first blood" was a way to skirt death
and end a duel civilly.

First blood has also come to be identified with the
idea of being the first to aggrieve another, or the first
to be aggrieved. If someone has drawn your blood first,
you have licence to return the violence — often dish-
ing out far more than you received. In the 1982 action
film *First Blood*, Sylvester Stallone — acting as Rambo —
tries to contain his emotions after being traumatized as

an American soldier captured by enemy forces in the Vietnam War. When the movie begins, he is back at home in the United States, drifting and trying to stay out of trouble. Pent up with rage and frustration, Rambo keeps it together until he is picked on by a sheriff. Rambo keeps a lid on his emotion, and displays no sign of violence until the sheriff draws first blood. After that, it's mayhem. Rambo carries out every manner of assault, death, and disorder, but the audience is led to root for him — and to accept the spectacle of violence that Rambo leaves in his wake — because his blood was drawn first.

We even attribute honour to animals that shed blood for our entertainment. In the bullfight, before the matador prepares to plunge his sword into the charging bull, the animal has already been weakened by repeated attacks from picadors. These armed men stab the bull while on horseback, making it bleed heavily before the final charge. In the eye of the crowd, a noble bull will maintain its fury and aggression even after suffering attacks that have left large quantities of blood to stream down its ribs. Ideally, the animal will be weakened in body, but not in spirit. The noble bull mounts a final charge, desperate to gore the lithe, nimble man with the red cape.

Audiences that watch cockfights seem to appreciate the very same thing: the undiminished, murderous intent of the cock, which charges forward and fights to the death, even as it is bleeding and being dismembered by a superior opponent.

Other animals have been made to fight to the death too, for human entertainment: dogfights are popular in much of the western hemisphere, as well as in Pakistan, India, and Japan. Some people pay to watch pit bulls rip apart feral hogs. In bear-baiting, the bears often have their teeth removed and claws filed before being matched against fighting dogs. Jack London's *The Call of the Wild* is narrated from the point of view of the dog Buck, who survives an attack by another dog and becomes an alpha male in a pack of sled dogs during the Yukon gold rush. Sled dogs then and now will rip each other to shreds if they spot a chance to pounce on another's weakness and rise in the hierarchy of the pack, thus giving rise to the expression "it's a dog-eat-dog world."

THE THRILL OF WATCHING beasts fight with nobility, even as they bleed, transfers into the human arena. I can think of no other professional team sport so obsessed with violence, and in which violence has such an overtly sanctioned role, as professional hockey. True, fighting will earn you a penalty. But it's part of the game, and fighting — along with the price of the penalty — is paradoxically used as a means to discipline players for displaying inappropriate aggression. Hockey teams even hire goons, whose primary role on the ice is to protect the more vulnerable members of their teams — the goalies and the high-scoring forwards — by fighting with those who dare to hurt them. The noise can be deafening in a hockey arena, but at no time is it more deafening than when two players break into a fight. The roar of the crowd makes me think of spectators at a gladiatorial duel in ancient Rome — but more on that later. Fighting is not countenanced in baseball, football, or basketball, but a hockey game without a fight will disappoint a diehard traditional fan.

Much has been written about the physical and psychological damage sustained by young men who are cast into the roles of hockey goons. In the summer of 2011, three former National Hockey League enforcers died: one from mixing painkillers with alcohol, one from suicide, and another from likely suicide. Despite the obvious dangers to the fighters, many hockey journalists, fans, coaches, and players continue to insist on the necessity of bloodshed. They claim that the threat of this violence prevents other violence from occurring on the ice.

Like duelling, fighting in hockey is highly ritualistic. Informal but very real rules are followed: each combatant should agree to fight, face the other directly, and drop his gloves; and it should end when the first fighter tumbles to the ice. Most fans and players accept fighting, and the misdemeanour usually warrants nothing more than a five-minute penalty. I cannot for the life of me understand why we allow men to brutalize each other like this in sport, and I do not believe for a moment that it is good for these combatants — in the moment, or later, when coping with head trauma and emotional stress.

Boxing and ultimate fighting (or fighting in cages) also stand out as some of the most bloodthirsty sports today. There are rules to be followed. You can throw a punch so hard that it kills your opponent — literally knocks him dead in the moment. That is within the rules. A knockout, or hitting an opponent so hard that he falls down and cannot get back up quickly and competently, is the ultimate and most manly way to win a boxing match. If you kill your opponent in the process, well, as long as it was a clean hit, it's one of the risks of the sport. Many boxers have died after taking excessive punishment in the ring. In 1982, a twenty-three-year-old South Korean boxer named Kim Duk-Koo slipped into a coma minutes after he lost a technical knockout to Ray Mancini at Caesars Palace in Las Vegas. Kim, who had grown up poor and been a shoeshine boy before becoming a professional boxer at the age of nineteen, died four days after the bout. As well, in a heavyweight boxing match in the Manchester arena in Calgary in 1913, Canadian brawler Arthur Pelkey knocked out his Nebraska opponent Luther McCarty in the first round. McCarty, who had been dubbed the next "Great White Hope," died in the ring. The Royal Northwest Mounted Police charged Pelkey with manslaughter. Jurors acquitted the boxer and shook his hand after rendering their verdict, but Pelkey had been traumatized by the event and never won another fight.

Boxing is highly ritualistic too. The combatants salute each other before the fight begins. Their gloves are monitored and controlled for size and makeup. They wear mouthguards. We allow one licensed boxer to draw blood from another. We even allow for the shedding of much blood, provided that it doesn't hinder a boxer's sight. You can aim for a bloody spot and try to draw out more red stuff. You can hit a weaker boxer over and over to draw out the punishment. This brings to mind Muhammad Ali, calling out, "What's my name?" as he spent fifteen rounds beating the stuffing out of Ernie Terrell in 1967, to avenge Terrell's insistence on calling him Cassius Clay—the name he had before becoming a Muslim. You can coax the blood to flow and you can beat your opponent so badly that he is thoroughly humiliated as a form of public entertainment, but you cannot hit a boxer when he is down. That's outside the rules. Not permitted by ritual. They stop every three minutes for a break so someone in their corner can try to staunch their bloody wounds. If blood is streaming down the face of a fighter, the fight goes on unless the bleeding fighter begins losing badly—perhaps because he cannot see his opponent properly—in which case the fight ends with a technical knockout.

In the year 1997, boxing fans witnessed what later became known as the "bite fight"—one of the most memorable fights in the past two decades. Boxers Mike Tyson and Evander Holyfield squared off in Las Vegas, Nevada. The previous year, Holyfield had won—by TKO—a match in which Tyson had been heavily favoured. When they met again, Tyson began once again to lose badly. In the third round, he bit off a chunk of Holyfield's ear—a big enough piece to spit onto the floor. Several minutes later, Tyson did it again. Shortly thereafter, Tyson was disqualified. He was also fined and had his licence revoked for a year. I'm not countenancing Tyson's behaviour. However, I do find it fascinating that the "bite fight" garnered such huge attention, in a sport where combatants — generally young black men from poor communities, who are watched by fans who pay thousands of dollars to attend championship fights — are regularly knocked out and concussed. Some sustain serious brain damage. Some die right there on the sweat- and blood-soaked floor, and others die far sooner than they should. It is a blood sport. And we love it.

Boxing has suffered in public popularity of late, because it is insufficiently wild and violent in relation to ultimate fighting. Fought in cages, using mixed martial arts, this sport seems more intense, savage, and varied. You can punch *and* kick. You can attack almost any part of the body. There are far fewer reasons for a referee to intervene, slow the pace of bloodshed, or impede blood lust. I'm betting that as time goes on, we will find even more savage ways to allow men — and women — to draw each other's blood and batter each other senseless.

WITH RESPECT TO CAUSING BLOODSHED, ritual forms the dividing line between criminal and honourable behaviour. We license the spilling of animal blood every day in most parts of the planet, but rules govern us every step of the way. In most wealthy nations, for example, you can no longer walk into a meat store and ask to have a chicken (or a larger animal) butchered before your eyes. For reasons of public health, and perhaps to accommodate public squeamishness, animals are to be killed and bled in slaughterhouses. This is meant to be carried out behind closed doors, and it is also designed to minimize the spread of bacteria to protect consumer safety. You will not be alarmed or surprised to see a truckload of pigs travelling at 110 kilometres an hour on the highway. It's a common sight. You know where they are going. But if the driver stopped to slaughter and roast a pig at a roadside picnic stand, it wouldn't take long for someone to call 911. Bloodshed is permitted. Bloodshed is necessary. But to win public sanction, it must follow an intricate set of rules and rituals.

Certain forms of bloodshed are not permitted, and warrant the maximum penalty imaginable. If you shoot and kill someone in a state where capital punishment is allowed — in Texas, for example — you stand a chance of being executed. We are back to Biblical references — many people these days rely on the dictum "an eye for an eye, a tooth for a tooth." If you bomb runners and bystanders at the 2013 Boston Marathon, taking three lives and spilling the blood of numerous other victims all over the street, many will clamour for your death, and the American government may take the unusual step of calling for the death penalty, even in a state that does not have capital punishment on the books. Bloodshed is permitted only in highly controlled rituals of publicly sanctioned violence. In such moments, we drink it up.

The regulated, ritualized, bizarrely accepted blood-shed haunts me all the more because of its contradic-tions. The English gentleman hunter on his horse is taken as the epitome of class, civility, and good breed-ing. Yet his joy consists of encouraging trained hounds to rip the stuffing from foxes. A gentleman keeps his salad plate on the left and his wineglass to the right. He marries well. He ensures that his children learn to read and write. But on Sundays, he indulges his love of fox blood.

Bloodhounds pursue more than red-coated foxes. They also carry the scent of humans in their noses when they famously pursue fugitive slaves and criminals. References exist to bloodhounds chasing men in medi-eval Scotland. In 1673, the Irish (or English, possibly) philosopher and chemist Robert Boyle wrote about an incident in which a dog followed a man's path over seven miles before locating him in a house.

The dog is, well, dogged in its pursuit of foxes and humans alike. In lore, the best way to shake him off your trail is to drag your body — blood included — through water. It takes one vital fluid to mask the scent of another.

The idea of dogs or monsters locking on to the scent of humans and then pursuing them with murderous intent is a terrifying thing to imagine. What avid child reader has not been captivated by the story of the giant's blood lust in "Jack and the Beanstalk"? In this British fairy tale, the giant is unsettled when he enters his house in the clouds and detects the scent of human blood. Jack, as we know, has climbed the beanstalk and has been asking the giant's wife for food.

The giant, sensing the story with his nose alone, roars:

Fee-fi-fo-fum!
I smell the blood of an Englishman,
Be he alive, or be he dead,
I'll grind his bones to make my bread

Blood, alas, can give us away by its very scent. This serves as a warning to child readers, and to adults who can extrapolate and imagine the senseless assault and butchery that has plagued mankind since the dawn of "civilization."

AS A YOUNG BOY, I was particularly frightened to read that the wolverine was such a nasty, bloodthirsty mammal that it would kill other animals for no reason. The biggest member of the weasel family, not generally exceeding fifteen kilograms in weight, this particular carnivore had a reputation for attacking much bigger animals. But the no-reason thing troubled me. Why would an animal rip another to shreds for the fun of it? I doubt that it is true that the wolverine acts, or kills, merely for its own entertainment. But the image of killing for its own sake haunted me.

Others too have been upset to the very core by the idea of senseless killing. *In Cold Blood*, the 1966 account by the American Truman Capote, explores the true story of two ex-convicts who invade the home of a wealthy Kansas wheat farmer to steal his money. Not finding any, they slit the throat of the farmer, shoot him in the head, and go on to murder his wife and two of their children. The crime, which took place in 1959, shocked and riveted Americans, and Capote's subsequent account became a bestseller. The bloodshed that seems irrational and inexplicable is the kind that frightens us the most.

Another form of murder sits much more easily with us: killing for entertainment. This is the central theme of Suzanne Collins's bestselling novel *The Hunger Games*. The story's sixteen-year-old protagonist, Katniss Everdeen, like most citizens in the post-apocalyptic nation of Panem, is subjugated by the ruling class of people in the capital city. Every year, as punishment for having once risen up against the ruling powers, each of the twelve subjugated provinces must offer up one boy and one girl, between the ages of twelve and eighteen, for a ritualistic bloodbath in the woods. The fight to the death is televised for the entertainment of the people in the Capitol, who are caught up annually in the orgiastic pleasure of viewing the participants, finding ways to assist or hinder them during the battle, and salivating as they die in the most hideous, gruesome ways. Only one survivor can normally win. Since the novel features a sixteen-year-old who has volunteered for battle to save her younger sister from having to do so, it is not hard to imagine how Katniss will fare against her opponents, many of whom are bigger, stronger, better trained, better equipped, and more bloodthirsty. *The Hunger Games* is an adventure story, but it can also be read as a condemnation of a society that lusts to see blood spill.

The novel indicts reality television for its exploitation of human suffering, and offers a futuristic version of the *munera* — gladiatorial combats — witnessed by the masses from the comfort of their gathering-places. As early as the third century BCE, thousands of Romans began to congregate in arenas for these spectacles. Prisoners of war, criminals, Jews, slaves, women, and others were offered up for the amusement of those who would come to watch, and cheer, and perhaps roar in favour of sparing the lives of those who fought valiantly in defeat. The spectacles were enacted to confirm the rulers' power, to warn others about the dangers of opposing the state, and to entertain Romans. The thrill was about the spill, but not all were seduced by it. The Roman philosopher and orator Cicero, for example, wrote about the Pompeii Games of 55 BCE: "But what pleasure can it possibly be to a man of culture, when either a puny human being is mangled by a most powerful human beast, or a splendid beast is transfixed with a hunting spear..."

Humans do like the sight of blood. We don't permit murder, but we certainly have no problem planting the idea in the public psyche. At a National Rifle Association conference in Houston in May 2013, delegates visiting the vendor displays could see a life-sized, three-dimensional, female target for sale. It was called "the Ex." When "the Ex" is shot, she bleeds.

It is not difficult to imagine why the National Rifle Association has had such an enduring influence in the United States, and has been so successful at opposing efforts to curb the use of guns by American citizens. The Second Amendment to the American Constitution invokes "the right of the people to keep and bear arms," although the specific meaning of those words is contested. People who believe that citizens should have the right to carry guns argue that the Constitution backs them up. Others, however, say that the Second Amendment does not offer a blanket right to all civilians, but rather to those who form part of a state-organized militia.

Regardless of constitutional interpretation, violence does beget violence, which is exactly what underpins some objections to capital punishment. With its long history of state-sanctioned shooting, hanging, gassing, electrocution, and lethal injection, the United States is the only G8 country that continues to carry out capital punishment.

For a country where bloodshed was wantonly and publicly celebrated as part of capital punishment, though, one must look to France and its use of the guillotine.

During the French Revolution, a violent movement during the last decade of the eighteenth century to overthrow the monarchy and institute republican government, the use of the guillotine rose as part of the so-called Reign of Terror. In less than one year (1793–94), it is estimated that more than sixteen thousand enemies of the revolution were guillotined in France. That works out to forty-three beheadings a day. The guillotine symbolized the Reign of Terror. Maximilien de Robespierre, one of the key revolutionary leaders, stated that "terror is nothing else than justice, prompt, severe, inflexible." Robespierre, who had successfully advocated for the guillotining of King Louis xvi in 1793, and who as part of the so-called "Committee of Public Safety" had vastly increased the number of public executions, eventually fell out of favour with his co-revolutionaries and was guillotined himself, in 1794.

Among the many victims of the Reign of Terror was Louis xvi's wife, Marie Antoinette, who was guillotined in Paris on October 16, 1793. Asked at her trial if she had anything to say about all the allegations against her, Marie Antoinette said, "I was a queen, and you took away my crown; a wife, and you killed my husband; a mother, and you deprived me of my children. My blood alone remains: take it, but do not make me suffer long."

Marie Antoinette spoke with the same dignity and composure as her late husband, Louise xvi, whose last words before being executed were: "People, I die innocent! Messieurs, I am innocent of all I am accused of. I hope that my blood may cement the happiness of the French people."

Marie Antoinette was thirty-seven years old on the day she was beheaded, some nine months after her husband was killed in the very same way. Louis XVI was given the opportunity to prepare for his trial, as well as a goodbye dinner with his family. Marie Antoinette had been detained in a dungeon at length and had no such privileges before she was hauled through the streets in an open cart en route to meet her executioner at a scaffold in the Place de la Révolution. She had been persecuted in a sham trial over two endless days (sixteen hours one day, and fifteen hours the next), and accused of charges including but going far beyond treason. Today, it is widely believed that the allegations were bogus, but the dethroned queen was also vilified for aberrant sexual behaviour.

For years, pamphlets featuring pornographic drawings of her had circulated widely in France, alleging that Marie Antoinette — still seen as a foreigner in France, given her Austrian heritage — had a rapacious sexual appetite for men and women. They alleged she organized orgies in the palace at Versailles. (Words that might have evoked pride among male aristocrats helped hasten the execution of the dethroned queen.) In trial, Marie Antoinette was also accused of sexually abusing her own son. When pressed for an answer to that charge, she refused to discuss the matter and simply said: "If I haven't answered it is because Nature herself refuses to answer such a charge against a mother. I appeal to all the mothers present — is it true?" While Marie Antoinette stood trial over two long days, she was bleeding profusely. Historians now speculate that she may have been suffering from uterine cancer. In a letter composed in her prison cell on the morning of her execution to her sister-in-law (who would never receive the letter, and who would soon also be guillotined), Marie Antoinette wrote: "I have just been sentenced to death, but not a shameful one, since this death is only shameful to criminals, whereas I am going to rejoin your brother." Among the countless thousands of people summarily executed by a regime gone wild with paranoia, the persecution of Marie Antoinette shows perhaps better than most the pitfalls of submitting to rulers who are so rabidly intent on extracting blood that the truth becomes nothing more than a stone to be kicked from view.

Although the rate of public executions fell off as the Reign of Terror came to a close, guillotining remained a fact of life in France as the standard mode of execution. The last people to be guillotined in Paris were Claude Buffet and Roger Bontems in 1972, and the last person executed in France was Hamida Djandoubi in 1977.

Considered barbaric by today's standards, the guillotine was named after Joseph-Ignace Guillotin, who had urged the French parliament to consider a more civilized and painless manner of beheading people. Previously, aristocrats had been executed by means of the broadsword or axe, but others were burned at the stake or broken over the wheel, or faced the long, slow strangulation of the gallows. During the Reign of Terror, Charles-Louis Sanson became the official executioner and praised the guillotine. On April 25, 1792, he said: "Today the machine invented for the purpose of decapitating criminals sentenced to death will be put to work for the first time. Relative to the methods of execution practiced heretofore, this machine has several advantages. It is less repugnant: no man's hands will be tainted with the blood of his fellow being..."

Quite apart from debates about whether the guillotine was faster, more painless, and more humane than other forms of execution in vogue at the time, beheading became a public spectacle during the Reign of Terror. Crowds attended in large numbers, some cheering and jeering, others mumbling that the executions happened too fast to allow for much enjoyment of the actual act. As Robert Frederick Opie notes in his book *Guillotine: The Timbers of Justice*, gallons of blood were spilled at the beheadings, and packs of dogs would come to lap it up at night at the base of the scaffold. The audiences cheered the beheadings as a form of entertainment, despite the quantities of blood spilled and the resulting stench. The revolutionaries who sentenced their enemies to death were so pleased with their new instrument of decapitation — and so intent on asserting their power and intimidating French citizens — that in one case, after a man who had been sentenced to death by decapitation managed to commit suicide first, his corpse was hauled to the scaffold, where it was guillotined anyway.

People respected the executioner, but they did not like him. Indeed, the role of public executioner — like many other trades — was assigned to specific family dynasties.

Sons and grandsons were trained in the work of their ancestors. It was a living, but it could not have been a happy or comfortable life for the executioner or his family. Opie writes: "Having entered into the profession, a family and his descendants were marked forever. There was no escaping their role; down through the generations brothers and sons were bound to the same employment and had to face the same prejudice. The only friends, associates and potential spouses of an executioner were the families of other executioners."

I have no difficulty believing Robert Opie. If I were living during the Reign of Terror, the very last person I'd be inviting over would be the executioner. Who wants a figure of death haunting your breakfast table and breathing over your baguette? I'd sooner go hungry than break bread with the man who guillotined my neighbour.

Alas, for the hundreds of years during which beheadings and other forms of capital punishment were carried out routinely in France, people seemed to enjoy the spectacle but they shunned the executioner. Looking back at this time in history, I try to imagine the courage of individuals who opposed the will and the control of the state. Simply living individualistically, or insisting on the right to express political beliefs, would have been enough to expose one's neck to the falling blade. In revolutionary France, and in so many other homicidal regimes, leaders used the spectacle of blood to exert power and to exact conformity.

IN THE PAST FIFTEEN YEARS, we have witnessed the publication of two series of books with unprecedented sales. Both series were aimed at children or young adults. And both dealt with blood. Blood didn't merely spill in the books. These days, it's hard to find a book or movie where blood *doesn't* spill. These books — which, combined, have sold well over half a billion copies and have been devoured by children (and many adults) worldwide — turn on the very concept of blood purity. At the very least, hundreds of millions of children in the past decade and a half have been swept away by books in which the characters are fixated on the quality of blood. The books have become so omnipresent that it seems almost redundant to mention their names: the Harry Potter series about wizards, by J. K. Rowling, and the Twilight series about vampires, by Stephenie Meyer.

In *Twilight*, Bella Swan is a "good girl" who falls in love with the vampire Edward Cullen, who, in turn, protects and falls in love with her. Edward insists that she must maintain her virginity by not making love with him until they are married. Edward, because he loves her, must transcend his instincts and help Bella maintain her sexual purity and the purity of her human blood by resisting the very thing that vampires just love to do to young girls: seduce them and suck their blood and turn them into undead who get to preserve their youth and vigour forever.

You would have to be brain-dead, these days, to fail to observe how commercially successful vampires have become in contemporary culture. True, people have been talking about vampires for many centuries. In 1897, Bram Stoker's novel *Dracula* helped set in motion contemporary obsessions with the undead who keep their youth and their powers (including legendary sexual prowess) intact by feasting on human blood. But in recent years, the focus on vampires seems almost a prerequisite for commercial success in the fantasy genre. *The Vampire Diaries*, *True Blood*, and *Buffy the Vampire Slayer* are just three in a litany of TV shows on the theme.

Is it because vampires are unbelievably sexy, drawing upon centuries of sexual experience each time they bring a young thing to the very peak of pleasure (and remaining wrinkle-free despite their advanced age)? Is it because vampires are all-powerful, and can easily rip out a heart or pull off a limb? Is it the way that a vampire corrupts humans, draws them over to the dark side by contaminating their blood — the ultimate sacred fluid?

Just as many young people are drawn to vampire culture, many are also drawn to cutting themselves as a form of controlled self-abuse. Experts theorize that cutting among young girls is not generally the expression of suicidal impulses, but rather a way of managing pain and anxiety. The vampiric seduction is a private act, as is the act of drawing out one's blood. People tend to get over their vampiric obsessions as they emerge from adolescence, as do most girls who have been drawn to cutting.

The vampiric attack is irreversible. Once you've gone over to the dark side, there is no coming back. You do get to live forever, but no longer as a human. Cutting, however, allows for more control. Who will see the marks, which you can cover up with clothing? How seriously are you to be hurt, by losing a little blood? For some, perhaps, cutting focuses one's pain in the body, instead of in the psyche. But it is temporary. And most adolescents grow out of it.

THE ONE RECENT CHILDREN's literary phenomenon even more famous than Twilight is J. K. Rowling's seven-part series about wizards, named after protagonist Harry Potter. Harry's life would have been a lot more peaceful if he could have taken easy refuge from a murderous, powerful, evil wizard named Lord Voldemort. Voldemort, the antagonist who pursues Harry to the very end of the series, is obsessed by notions of blood purity. Indeed, it is possible to read the Harry Potter series as a meditation about good versus evil, along the lines of blood. In Harry's world, there are wizards (or witches) who descend only from other wizards, and others who have one muggle (human) parent and one wizard parent. There are even muggles who become wizards and can wield the power of wizards, without having a parent who is a wizard. Harry himself is the son of a wizard (father) and muggle (mother). Even though almost nobody in the world of Harry Potter is a pure-blooded wizard, Voldemort (also of mixed background) and his followers are on a vendetta to exterminate wizards with any so-called blood impurities, who are known as "mudbloods" or "half-bloods." Draco Malfoy, for example, is a wizarding student who sides with Voldemort, considers himself a pure-blood, and insults Harry's friend Hermione Granger by calling her "a filthy mudblood."

Ron Weasley, a wizard who becomes close to Harry and Hermione, explains to them early in the series: "Mudblood's a really foul name for someone who is Muggle-born — you know, non-magic parents. There are some wizards — like Malfoy's family — who think they're better than everyone else because they're what people call pure-blood...I mean, the rest of us know it doesn't make any difference at all."

The struggle between good and evil in the world of Harry Potter comes down to those (such as Voldemort, Draco, and his father, Lucius) who would exterminate people of impure blood, and those (such as Harry, Ron, Hermione, and their beloved wizarding school headmaster, Albus Dumbledore) who wish to stop the exterminators and live in peace.

J. K. Rowling has noted that in writing the Harry Potter series, she had in mind the obsessions with racial purity held by Nazis and other white supremacists. In answer to the question "Why are some people in the wizarding world called 'half-blood' even though both their parents were magical?" she has said:

The expressions "pure-blood," "half-blood" and "Muggle-born" have been coined by people to whom these distinctions matter, and express their

originators' prejudice. As far as somebody like Lucius
Malfoy is concerned, for instance, a Muggle-born is as
"bad" as a Muggle. Therefore Harry would be considered only "half" wizard, because of his mother's parents. If you think this is far-fetched, look at some of the
real charts the Nazis used to show what constituted
"Aryan" or "Jewish" blood. I saw one in the Holocaust
Museum in Washington when I had already devised
the "pure-blood," "half-blood" and "Muggle-born" definitions, and was chilled to see the Nazis used precisely
the same warped logic as the Death Eaters. A single
Jewish grandparent "polluted" the blood, according to
their propaganda.

The books in the Harry Potter series have sold more
than 450 million copies. They are among the most influential and widely read — and maniacally adored — children's books ever published. When my eldest daughter,
Geneviève, was ten years old, I took her to see J. K.
Rowling read at the SkyDome sports arena in Toronto.
Thousands of children were in attendance. It was the
biggest literary reading I have ever attended. Every child
I looked at in the audience seemed not only enthralled
but also thoroughly acquainted with every single word
the author read aloud. It still astounds me to think that
one of the most famous, bestselling books in the history
of human civilization speaks at great length to children
about blood and identity.

THE OBSESSION WITH BLOOD PURITY is imaginary in the Harry Potter series, but it underpins murderous tendencies and genocidal behaviour that have repeated themselves all too many times in real life. Over and over, in the course of history, humans have invoked notions of blood purity to justify atrocities.

Perched in her seventeenth-century convent in Mexico City, Sor Juana expressed a fear of writing a spiritual analysis that might offend the Holy Office, the Catholic body responsible at the time for carrying out persecutions against Jews and other people deemed heretics in Europe as well as in New Spain. "I want no trouble with the Holy Office," she wrote, "for I am but ignorant and tremble lest I utter some ill-sounding proposition or twist the true meaning of some passage."

The Spanish Inquisition, which began in the 1400s on the Iberian peninsula (Spain and Portugal), crossed over the Atlantic and continued in the New World in the seventeenth century. Sor Juana would have been well informed about the thousands of Jews, Moors, and others who two centuries earlier had been persecuted, tortured, and — if they hadn't already been burned at the stake or murdered in other ways — expelled from Spain.

Centuries before the Spanish Inquisition, early persecutors often justified the murder of Jews by falsely claiming that they had killed innocent Christians — generally young boys — and mixed their blood with unleavened bread during Passover rituals. This allegation against Jews came to be known as "blood libel."

Judaism explicitly forbids the consumption of blood. In Leviticus 17:10, God plainly instructs Moses to this effect: "And whatsoever man there be of the house of Israel, or of the strangers that sojourn among you, that eateth any manner of blood; I will even set my face against that soul that eateth blood, and will cut him off from among his people." In addition to barring the direct consumption of blood, Leviticus clearly spells out that blood is to be removed from meat before it is eaten. Jewish dietary laws are replete with rules about blood and food, including the stipulation that one must kill an animal humanely by slitting its throat with a sharp knife and draining the blood quickly. Despite these clear prohibitions, blood libel arises in early medieval times as a pretext to attack Jews.

The first documented case of blood libel involved the accusation that Jews in Norwich, England, had killed, by means of crucifixion, a twelve-year-old apprentice tanner named William. The accusations were never proven, but they grew and multiplied and led to additional accusations that Jews were preying on Christian boys to carry out ritualized murder. This led to the slaughter of many Jews in England, and to their eventual expulsion in 1290. (Many resettled in Spain, from which their descendants in turn were expelled, two centuries later.)

Muslims invaded the Iberian peninsula in the year 711, controlling diminishing portions of the region for some seven centuries, until the Catholics finally drove them out at the end of the Reconquista in 1492. In the

years between, as the Catholic royalty, church, and followers fought to reclaim the land they believed to be
theirs, Jews — who had been accepted to varying degrees
for centuries by the Muslims — began to be vilified,
restricted in their civil rights, attacked, and murdered.
Priests and their followers began murdering Jews in
Seville and elsewhere in 1391. Some Jews were enslaved,
others fled, and still others converted to Christianity in
a bid to escape persecution and death. These converted
Jews became known as the *conversos*. There were more
Jews in Spain than anywhere else in the world at the
time, and it is estimated that more than one hundred
thousand became *conversos* — many of them making
sincere conversions that passed down through the generations. Being a *converso* gave a former Jew (or a person
descended from former Jews) full rights of citizenship,
and many of them thrived in business, government,
church, moneylending, and other disciplines. They also
married into Christian families, nobility included.

In 1449, however, Spanish citizens began to riot
against *converso* tax collectors in Toledo. Authorities
issued laws requiring the *conversos* to demonstrate the
purity of their blood (*limpieza de sangre*, as it is known in
Spanish). In the first step toward openly suggesting that
Jews were biologically distinct from Christians, authorities issued a law requiring that the *conversos* demonstrate
their own blood purity. As Erna Paris writes in *The End
of Days: A Story of Tolerance, Tyranny, and the Expulsion of
the Jews from Spain*, "Once *conversos* had been effectively

reidentified as Jews by the statute of exclusion, all the ancient anti-Jewish accusations could be revived with impunity... 'Purity of blood,' or *limpieza* (cleanliness), exploded into a national obsession."

Within half a century, thousands of Jews and *conversos* were deprived of their rights, attacked, tortured, burned at the stake, and killed by other means. By 1490, some 2,000 conversos had been burned to death during or immediately after popular public events known as *autos-da-fé* (Portuguese for "acts of faith"). Finally, at the end of the Reconquista, in 1492, the Catholic monarchs — King Ferdinand and Queen Isabella — drove the Muslims and an estimated 150,000 to 200,000 Jews from Spain. Tens of thousands of Jews died trying to flee the country, according to the Jewish Virtual Library (JVL). Many of the most unlucky refugees ended up in Portugal, from which they were expelled again a few years later. Some of the more fortunate migrants ended up in Turkey. As for their fate, the JVL says: "Sultan Bajazet welcomed them warmly. 'How can you call Ferdinand of Aragon a wise king,' he was fond of asking, 'the same Ferdinand who impoverished his own land and enriched ours?'"

The obsession with blood purity robbed the Iberian peninsula of many of its most talented, productive, educated citizens. The Inquisition, or persecution of heretics, plunged Spain into the Dark Ages and raged on for centuries. Erna Paris notes that after the Reconquista, pure blood became a condition for every post of merit: "By 1673, a 'Jew' was being described as someone with as

little as twenty-one degrees of blood relationship, or …
(as) an Old Christian who had been suckled by a wet
nurse of 'infected blood.' The word Jew was stripped of all
content; like 'pure' and 'impure,' it was a multipurpose,
abstract trigger for class hatred, rejection and otherness."

In Spain, to this day, the city of Santiago de
Compostela — site of one of the most significant religious meccas in the world — is named after the Apostle
James, whose remains are said to be buried under the
cathedral in the old town. Catholic legend has it that
Santiago rose from the dead to inspire Catholic warriors
to victory over the Muslims in Clavijo in the year 844,
when the Spaniards had embarked on their centuries-
long Reconquista to recapture the Iberian peninsula.
Henceforth, Santiago was much more than the simple
name of an Apostle. It hollered out the notion of drawing blood from the infidel. Since the battle at Clavijo,
the saint has been known throughout Spain and celebrated prominently in Spain's most famous cathedral
as Santiago Matamoros, which means "Saint James the
Moorslayer."

OBSESSIONS WITH BLOOD — as demonstrated by the persecution and murder of Jews in medieval Europe — have
spilled into countless acts of ethnic or racial hatred over
the centuries. I have examined the Spanish Inquisition
not because it is the only major example of human mon-
strosity — we all know that it isn't — but because the
same blood-based obsessions found in medieval Spain

spread into other countries and continents as new forms of genocide took root in the world.

Perpetrators of genocide — who massacred indigenous peoples in the Americas, the Armenians in the Ottoman Empire (now Turkey), the Jews during the Holocaust, ethnic minorities in Cambodia, the Tutsi in Rwanda, the people of Darfur, the people of the Democratic Republic of the Congo, and so many others — all demonized their victims by alluding, directly or indirectly, to the impurity of their blood. In the case of the Rwandan genocide, which took the lives of an estimated eight hundred thousand Tutsis and moderate Hutus in 1994, radio broadcasts and newspaper reports repeatedly referred to the Tutsis as "cockroaches" and urged listeners to kill them. Years earlier, the Belgian colonizers of Rwanda had helped drive a wedge between the two peoples by issuing identity cards to each group, and by artificially attributing to the Tutsis a role of economic and social privilege. Over the decades, it became possible for one group to demonize the other, and for the perpetrators of the genocide to successfully categorize the Tutsis as members of an inferior and threatening race, who deserved to be exterminated. Even the genocide in Rwanda rode on a racist wave, and was kept afloat by the implicit assertion that the Tutsis had tainted blood.

It has become common to hear genocidal massacres depicted as "ethnic cleansing," but I find the term repugnant. However unintentional, the language we use to describe abhorrent behaviour tends to reinforce the behaviour itself. How can there be such a thing as "ethnic cleansing"? How does murder, or genocidal massacre, relate to cleansing? The term is worse than a euphemism because it makes it sound as if it is actually possible to "clean" a society by ridding it of people of a certain ethnicity. I am aware that "ethnic cleansing" became a quick, shorthand, widely recognizable way for people to refer, for example, to the horrors of the civil war in the former Yugoslavia. I don't think that users of the term mean any harm by it. But the language that we use does affect the way we think about things and frame them. "Cleansing" sounds like a positive act. For me, there can be no cleansing in the context of deliberate bloodshed. There can only be murder.

The term *genocide* was coined in 1944 by the Polish-American lawyer Raphael Lemkin, who based it on the root words *genos* (Greek for "family" or "race") and *cide* (Latin for "killing"). Its catalysts are well known, although they still mystify us. The plague in medieval times led to pogroms against Jews, who were blamed for causing mass death by poisoning the wells of Christians. Genocide — the mass murder or extermination of targeted groups of people — has been taking place for as long as humans have existed. In his book *Blood and Soil: A World History of Genocide and Extermination from Sparta to Darfur,* Ben Kiernan notes that only recently have we come to roundly condemn genocidal practices. He notes archeological evidence of mass murder carried out against men, women, and children as many as seven thousand years ago in present-day Germany, and two thousand years ago in present-day France.

Kiernan reminds us that the Old Testament is replete with examples of genocidal enmity. Deuteronomy 20:17, for example, says: "But thou shalt utterly destroy them — the Hittites, Amorites, Canaanites, Perizzites, Hivites, and Jebusites — as the Lord has commanded you." The Qur'an enjoins believers to "slay the idolaters wherever you find them." Christians, Hindus, and others have perpetrated every manner of genocide.

No one group holds a monopoly on this form of ultimate evil, and few groups have completely avoided capitulating to it. But we abase ourselves each time we stand by passively and allow one group of humans to target another for destruction.

One of the most barbaric and effective ways to undermine an entire group of people has been to attack its women sexually, leaving them either dead or else "tainted" and socially isolated in the wake of the assaults. Rape has been a constant companion to genocide — during the Holocaust, over the course of the transatlantic slave trade, and in the former Yugoslavia, Rwanda, and the Congo, to name a few places. When it does not lead to their immediate death, it has been used to kill women slowly by means of infecting them with sexually transmitted infections and diseases such as HIV/AIDS. Failing literal death, it can bring about women's cultural death. Abandoned by husbands, shunned by those who consider the rape victims to be polluted, people who have been raped have often been marginalized in the very societies to which they belong.

Pointing solely to atrocities in other communities allows a person to have a false sense of moral superiority, so we must not flinch from acknowledging, opposing, and righting injustice in our own backyard. The Beothuk people were wiped out from Newfoundland. As a form of cultural genocide, thousands of First Nations children in Canada were forcibly removed from their families and shipped to residential schools, where they were abused and punished for showing any trace of their language or culture. Indigenous peoples were also slaughtered in the United States. In one lesser-known case of North American genocide, for a quarter-century ending in 1873 — with the sanction of law authorities and the cheering of newspapers — individuals, groups, and militiamen chased down and murdered nearly all of the Yana peoples of California. Atrocities have been carried out in most corners of the world, by many peoples of the world, and that they continue to this day. We must be ever vigilant. In our own backyards, and elsewhere.

NATURAL DISASTERS AND ECONOMIC anxieties bring out the worst in human beings. When we are insecure, when we feel that the outside world or the elements of nature threaten us, we look for a scapegoat. Often, that scapegoat has been a "privileged minority" that is seen to be usurping the rights and the dominant role of the majority. To find a scapegoat, the easiest path is to isolate a group of people — often along lines of their blood, or race. We fear that we will be displaced by these people, so we demonize them — right down to the level of judging the purity of their blood — and then we wipe them out. To protect ourselves. Xenophobia seems to be one of the most deeply entrenched human fears, and — aided by antiquated, nonsensical notions of blood — we humans have allowed it to bring out the very worst in us.

To recognize the fundamental equality of all human beings means that we cannot create hierarchies along lines of gender, race, religion, age, sexual orientation, or ability. To identify and shed subconscious beliefs that should be relegated to the Dark Ages, we must agree that blood is no determinant of human difference. In our bodies, and in the red stuff that courses through our veins and arteries, we are one and the same.

Blood has the ability to bring us together, when we share it to save each other's lives, or draw upon it metaphorically to allude to the most noble elements of the human heart and soul. But, sadly, our fixations on blood have all too often driven human beings apart; given us the most facile, unexamined, and absurd excuses to demonize each other; and fuelled the most atrocious behaviour known to mankind.

Blood. How about if we take it or leave it? Let us take it, when it is offered in the way of help — medical or metaphysical. And let us leave it right where it belongs — circulating in our veins and arteries, nothing more and nothing less — when it comes to dealing with our most base human instincts.

FIVE

OF PRESIDENTIAL MISTRESSES, HOLOCAUST SURVIVORS, AND LONG-LOST ANCESTORS: SECRETS IN OUR BLOOD

MANY FAMILIES HAVE SECRETS of the blood. Mine certainly does. My paternal grandparents, May Edwards Hill and Daniel Hill, had to elope to get married in 1918, shortly before he went overseas as an American army officer during World War I.

May came from a well-to-do Catholic black family, and there was nothing satisfactory about the Baltimore-raised young man with whom she had fallen in love. May's mother, Marie Coakley, who could pass for white but was married to a black dentist and was living as a member of the so-called "black bourgeoisie" in Washington, D.C., led the charge against my grandfather. What was so bad about him? In the opinion of my great-grandmother, Daniel Hill had three strikes against him. He was not Catholic. (He was soon to become a

lifetime minister of the African Methodist Episcopal
Church, and therefore close to being a heathen. Indeed,
when May wrote love letters to her husband while he
fought in the trenches of France, she began some of
them with "My dear pagan buzzard.") He was not from
money. And he was a dark-skinned African-American—
exactly the opposite of what Marie Coakley wanted for
her daughter. How would their children slide into white
society, or hang on the edges of it, or benefit by means of
association, with May married to a visibly black man? I
heard many stories about Marie Coakley trying to break
up my grandparents. I even dramatized and exagger-
ated the situation in my novel *Any Known Blood*. But
here, I will stick to the family legend as I have heard it.
According to the stories that have been passed down to
me, Marie offered to pay for May's tuition at Radcliffe
College only if she would leave *that man*. Daniel eventually moved his family to Missouri (where my father
was born, in 1923), then to Colorado, and then to Oregon,
partly to escape the clutches of his mother-in-law.

However, in 1919, before May and Dan moved out
west, Jeanne—the first of their four children—was
born. The war had just ended. At the time, May and Dan
were living in Philadelphia. After giving birth to Jeanne,
May was kept for a long time in the hospital. With May
still confined to her hospital bed, Marie Coakley made
arrangements for Jeanne to be baptized in a Catholic
church (with a white congregation, and a white priest)
in Philadelphia. The priest, the story goes, did not

recognize that Marie Coakley was black, and Marie did not tell him. Not long after May was released from the hospital, she, her mother, her husband, and baby Jeanne travelled to the church on the day of the baptism. When the priest saw that they were black, he refused to baptize Jeanne. "They thought they were showing up for a baptism, and they were all completely insulted," Doris Cochran, my aunt (and the only surviving member of the family), recalled in a telephone interview in June 2013. "The priest refused because Jeanne was a child of colour."

Jeanne, who died in 2005, told me (in 2000, when I was researching *Black Berry, Sweet Juice*) that she ended up being baptized in a black Catholic church in Philadelphia, but that the incident left her mother disenchanted with Catholicism. May eventually left the Catholic Church entirely. According to Jeanne, the Protestant churches in the early 1900s were just as racist. Nonetheless, Jeanne attended the Catholic Church until 1985, when she finally moved over to the African Methodist Episcopal Church.

On the day of our interview in her home in Brooklyn, Jeanne was full of stories of family secrets. She had reached the age of eighty-one and seemed to have lost any sense of urgency about keeping the secrets locked away. She said that her grandmother Marie Coakley was born after Jeanne's great-grandmother Maria Coakley was raped while working as a maid in the White House in 1875, during the presidency of Ulysses S. Grant. With a grin on her face, Jeanne said, "There was involuntary mixing in

the White House." Jeanne said that Maria Coakley neither
named nor accused her rapist. Maria gave birth to Marie
on February 21, 1876, and the baby Marie was raised as
one of the children of her grandparents, Jenny and Gabriel
Coakley. Jeanne recalled that Marie's "siblings" and "parents" were considerably darker than she was.

My Aunt Jeanne said she did not hear of this story
until she was twenty-one years old and happened to
be visiting with her grandmother Marie and with
her great-aunt Gertrude. Marie happened to insult
Gertrude. Gertrude shot back that Marie was a bastard,
and taunted her about being conceived as a result of rape
in the White House.

I have no way of verifying the story. I will never
know, for sure, if my great-great-grandmother was raped
in the White House. I have no way of knowing if one
of my ancestors on my father's side was a White House
employee or a visitor, and white to boot. I asked my Aunt
Jeanne for her opinion on this family story, which is a
secret no more. "It is probably true," Jeanne said. "Things
like that happened all the time." I know a few secrets
about my family, but I do not know them all.

Our blood contains many secrets. These secrets may
have to do with an identity — racial, religious, or other —
that we have chosen to shed, hide, or alter. They may
have to do with crimes we have committed. Children
we have fathered, or mothered. People to whom we are
related, or not. Blood also has the potential to yield up
secrets — and even resolve disputes — about our most

distant ancestry. Blood can be the most intimate reflection of our being, and it can offer details that are either abhorred or welcomed. Since blood and race have come to be so intimately acquainted, I will spend a good part of this chapter assessing that relationship. However, I also want to touch down on blood and crime, blood and forensic investigation, and how the science of genetics may be altering or challenging long-held assumptions about the meaning of our blood and the nature of our identities.

IT SO HAPPENS THAT BLOOD is one of the hardest stains to remove. Apparently, this is because hemoglobin binds with fibres, rather than just sitting on top of them (as, say, mud might do). You don't necessarily want your house guests to know that you have spilled blood on bedsheets or towels. Blood, in these cases, is something you want to keep secret.

For years, skilled laundresses have had all sorts of household tricks to remove bloodstains from fabrics. The first lesson is to avoid the use of hot water. No need for a washing machine, either, although time is of the essence. If you have spilled blood, you don't sit back and contemplate your fate. You act right away, sopping it up with something absorbent and using cold water. On the Web, you can find endless suggestions about how to deal with the problem. In fact, the website *Mrs. Clean* offers specific tips: cold water, soap, and hydrogen peroxide. Failing that, you may also use your own saliva or reach for toothpaste, meat tenderizer, or ammonia.

The strange thing about blood on sheets is how exposed it makes us feel if others see it. Saliva on a pillowcase, semen on a bedsheet, blueberry stains on a blanket — none of these things bring about quite the self-consciousness one feels about leaving blood on the sheets in the house of an old friend who has just given you dinner and put you up for the night. In most cases, we take steps to clean up after ourselves so nobody will see the signs of blood in our beds. But there is one situation in which the very opposite approach is required.

In chapter one, I mentioned that men don't have to think about monthly bleeding. Another thing that men don't have to worry about is proving their virginity by producing a bloodstained bedsheet after the wedding night — a ritual so ancient that it's difficult to trace its exact origins. The blood on the sheet is intended to come from the breaking of the hymen, a thin membrane that covers the opening of the vagina until it is broken during intercourse. Therefore, the thinking and the tradition go, blood on the wedding sheets proves the virginity and thus the purity of the bride, and should be hung out the following day to prove to the family and community that the groom got a good deal. Blood, in this case, is meant to put to rest any suspicions that the newly married woman has a secret sexual past.

There are several problems with this way of thinking. First of all, not all women are born with a hymen. Then there is the fact that the hymen can be broken by tampon use and physical activity. And there is the larger problem of judging chastity as such an essential asset for a woman that the absence of its proof can threaten the legality of the marriage and possibly put her in mortal danger.

Though the tradition of the wedding sheet is ancient, the value placed on virginity, and the presence of an intact hymen, persists today. And it has modern solutions. Hymen reconstruction surgery, in which a broken hymen is sutured or a piece of the vagina is used to create a new hymen, is available in many parts of the world. Some patients opt to create the illusion of chastity by having a capsule of coloured gelatin inserted into the vagina, so that it will break on the wedding night and create the desired "blood" stain.

This age-old blood revelation is meant to prove a woman's worthiness. It is a hideous way to evaluate a woman, exposing her to the judgement and condemnation of family and of the wider community. It is just one of the many ways in which society has used blood to expose our secrets, and it has led to fatal consequences for those who failed to satisfy their examiners.

LITERAL BLOODSTAINS HAVE some prominence in the mind, but the blood of the imagination can be far more powerful — it prevents us from hiding from ourselves the secrets of our crimes. Consider the emotional torture of Shakespeare's Lady Macbeth. In one of the most memorable meditations on blood in Western literature, Macbeth muses that an entire ocean couldn't rinse clean his bloody hands after he and Lady Macbeth conspired to murder King Duncan. And later, as Lady Macbeth wanders her castle, half mad, a doctor and a gentlewoman wonder why she has been rubbing and washing her hands incessantly. As they eavesdrop on her wild musings, Lady Macbeth incriminates herself with her own haunting memories of the murder. Referring to the victim, she says: "Yet who would have thought the old man to have had so much blood in him?" She is completely unable to rid her mind of the blood she has spilled. "Out, damn'd spot!" she cries. "Out, I say!" And a few lines later: "What, will these hands ne'er be clean?" She even fixates on the scent of her victim's blood, as evidence that refuses to go away: "Here's the smell of blood still: all the perfumes of Arabia will not sweeten this little hand."

Lady Macbeth cannot turn her mind away from blood, because more than anything else, it symbolizes the life she has just stolen from Duncan. She has blood on her hands, and she can never forget it.

Another powerful image of blood spilled on literary pages comes from Fyodor Dostoevsky's novel *Crime and Punishment,* in which the impoverished young man Raskolnikov murders a pawnbroker and her stepsister and then, in a state of mind similar to that of Lady Macbeth, virtually drives himself crazy with the memory of what he has done.

Let me share a few lines about Raskolnikov right after he commits the double murder:

His hands were sticky with blood. He dropped the axe with the blade in the water, snatched a piece of soap that lay in a broken saucer on the window, and began washing his hands in the bucket. When they were clean, he took out the axe, washed the blade and spent a long time, about three minutes, washing the wood where there were spots of blood, rubbing them with soap. Then he wiped it all with some linen that was hanging to dry on a line in the kitchen and then he was a long while attentively examining the axe at the window. There was no trace left on it, only the wood was still damp. He carefully hung the axe in the noose under his coat. Then as far as was possible, in the dim light in the kitchen, he looked over his overcoat, his trousers and his boots. At the first glance there seemed to be nothing but stains on the boots. He wetted the rag and rubbed the boots. But he knew he was not looking thoroughly, that there might be something quite noticeable that he was overlooking.

Like Lady Macbeth, Raskolnikov is never quite able to get past the sensation that the blood on his hands has, metaphorically, not washed away; hence his grisly act will haunt him ceaselessly. *Macbeth* and *Crime and Punishment* have endured in the collective imagination for their eerie portrayal of those who have carried out violent crime. We imagine the characters will never be able to wash from their minds the memory of blood spilled — on the floor, on bodies, on their own hands.

Blood is so red, and it stains with such unrelenting diligence — as liquid manifestation, and as enduring metaphor — that when we spill it criminally, we are forever stained. Maybe our blood knows this. Maybe we are filled with red blood for a reason. Maybe the nature, colour, and undeniability of our blood serve to help preserve mankind.

WHILE LADY MACBETH AND RASKOLNIKOV were unable to shake their minds free of the crimes they'd committed, today's criminals not only have to contend with their conscience (we imagine) but they must also deal with the new science of forensic examination. Blood spatter analysis can reconstruct the details of a crime to a forensic investigator. Did the blood drip slowly after a crime took place? Did it travel from the victim at medium velocity as the result of a blunt blow from, say, a bat or a fist? Did it leave the victim's body at high velocity, as the result of a bullet? If a bullet passed through a body, spatter analysis can reveal the blood that left the entry point and the blood that left the exit point. Investigating detectives will also look for voids in (or sudden absences of) bloodstains at a crime scene. Perhaps, when the blood sprayed, the assailant got some blood on himself (or herself). Perhaps the absence of blood in a certain spot suggests that a witness or an accomplice was at the scene of a crime.

Police officers who are interested in detective work will study bloodstain pattern analysis. By studying blood on a surface, it is possible, using trigonometry, to ascertain the angle at which the blood travelled and where it came from. Forensic examination can also reveal if a victim was shot at very close range: in that case, some of the victim's blood might even have been sucked back into the muzzle of the gun.

So evolved is the field of bloodstain analysis that there is even an organization called the International Association of Bloodstain Pattern Analysts. Thanks to modern science, we can tell if a bloodstain comes from a human or an animal. We can tell the blood type of the person who left the stain. The criminal may have washed clean the crime scene, but investigators can find traces of blood that are unapparent to the naked eye or are out of sight. Forensic analysis may well give the police many clues about who you are, what you used, and where you stood when you fired that gun or used that knife. And if they get a look at your own blood, they can extract information you may have kept secret: what you drank, what drugs you took, who your children and parents are, and what diseases lie hidden in your veins.

What the DNA analysis of blood tells us about a crime, however, is far from straightforward, or foolproof. It is open to interpretation. When O. J. Simpson stood trial for the 1994 murder of his ex-wife Nicole Brown Simpson and her friend Ronald Goldman, disputes about the reliability of DNA evidence extracted from blood samples came to be known as the "DNA Wars." Simpson's lawyers, arguing that the police had handled the blood evidence carelessly, succeeded in winning his acquittal in court.

(Two years later, however, a civil court in California found Simpson liable for the death of Goldman and the battery of Brown Simpson.)

In the end, it appears that the results of DNA testing of blood are similar to other forms of evidence introduced in criminal or civil courts: they may be used to acquit or convict. And the higher the stakes, the more likely the blood will be subject to competing analyses. Blood adds to the picture, and it complicates it, but it does not lessen the need for other forms of evidence.

BLOOD CAN CONTAIN MANY SECRETS. Some may never be revealed, and others may be exposed after lying dormant for decades or centuries. Given the endless obsession with separating people into artificially defined racial (or blood) groups, as outlined in chapter three, it is no wonder that thousands of individuals have attempted to "pass" into safer, unpersecuted groups.

In his memoir, *The Color of Water*, American essayist and saxophonist James McBride describes growing up black in the housing projects of Brooklyn, never having met his biological father, and having been raised with eleven siblings by a mother who he had always assumed was a light-skinned black woman. It was not until his adulthood that McBride discovered his mother had been born into an Orthodox Jewish family. After taking up with a black man, she had hidden the fact that she was Jewish and allowed neighbours, friends, and her own children to believe that she was a light-skinned black woman.

McBride recalls asking his mother whether God was black or white. He did not yet know his mother's family history, but had noticed her skin was lighter than his. She prevaricated brilliantly:

> "God's not black. He's not white. He's a spirit."
> "Does he like black or white people better?"
> "He loves all people. He's a spirit..."
> "What color is God's spirit?"
> "It doesn't have a color," she said. "God is the color of water. Water doesn't have a color."

Does this mean that James McBride's mother was indeed not black, during all those years that people assumed she was? What made her white or black, or Jewish, for that matter? We now know her secret, but for the longest time her fictional identity was "real" for the children who must have thought they knew her best.

It is possible — indeed, not at all rare — for a person to suppress or hide one identity and to offer another to the world. Sometimes, one does so as a matter of personal choice, as in the case of James McBride's mother. In other cases, passing out of one race or religion and into another can be a matter of life and death.

Just as those responsible for repressing and killing Jews and Muslims during the time of the Inquisition were fixated on the purity of Catholic blood, the same notions emerged during the Holocaust. Adolf Hitler's book *Mein Kampf,* which spews religious and racial hatred for nearly seven hundred pages, linked his so-called "Jewish menace" to what he considered to be the impurity of Jewish blood, the nobility of so-called Aryan blood, and the polluting dangers of miscegenation (procreation between people of different races). In the chapter entitled "Nation and Race," Hitler wrote: "Historical experience...shows with terrifying clarity that in every mingling of Aryan blood with that of lower peoples, the result was the end of the cultured people." Hitler's own words remind us how deeply genocide can be linked, in the mind of the perpetrator, to notions of blood.

It is no wonder that during the Holocaust, some Jews attempted to escape extermination — and some even succeeded — by adopting Christian identities. In 1999, Edith Hahn Beer published her memoir, *The Nazi Officer's Wife: How One Jewish Woman Survived the Holocaust*, which recounts how she grew up in a non-observant Jewish family in Austria but passed for a Christian to avoid being killed during the Holocaust. She married a Nazi party member named Werner Vetter, who became a wartime officer. He knew that she was Jewish, and when she was pregnant he often told her "that the Jewish race was stronger, that Jewish blood always dominated," but that he still looked forward to the arrival of the baby.

Hahn Beer kept her true identity hidden until the war's end. But she recounts a 1943 meeting with a Nazi official to obtain a marriage licence. For this, she had to prove that she was "German blooded." She had some false papers, but not all that she needed. When the official interrogated her about her maternal grand-mother, Hahn Beer lied and said that she was unable to obtain her grandmother's racial papers. The official scrutinized her, said it was obvious from looking at her that she "could not possibly be anything but a pure-blooded Aryan," and stamped a form asserting that she was "German blooded."

When Hahn Beer gave birth to her daughter, she refused sedatives in order to stay alert and avoid making any accidental reference to her Jewish identity. After the war, she had her daughter baptized as a Christian to sat-isfy her husband and ensure he would accept his daugh-ter. But he said the baptism had no effect on him because it was the child's "Jewish blood" that counted, and they soon divorced. Later, Hahn Beer moved to England, where she married a fellow Jewish Austrian. They lived together for nearly thirty years. After he died, Hahn Beer moved to Israel.

What if the Holocaust and World War II had con-tinued well beyond 1945? What if Edith Hahn Beer had remained a Christian to virtually all those who knew her, until she died naturally? How would people who knew nothing of her identity have identified her reli-gion? In a world where being seen as a Jew was a death sentence, would being perceived, treated, and even bur-ied as a Christian have made her a Christian? As long as her secret remained intact, she would have remained a Christian to everyone except herself.

Edith Hahn Beer was not alone in her strategy. The

United States Holocaust Memorial Museum's exhibition *Life in Shadows: Hidden Children and the Holocaust* notes that many Jewish children were baptized into Christianity, adopted Christian identities, and rigorously suppressed their Jewish ancestry in order to avoid extermination. Being divorced from their families, culture, and religion carried a steep emotional cost. But it saved their lives, and some were able to reintegrate with surviving family members after the war.

As the museum exhibition explains, many children had to wait until the end of the war to discover that their parents had been killed. Others were kept hidden from their parents after the war, and in some cases the families had to pay "redemption fees" to reclaim their children.

One of the children featured in the exhibition is Lida Kleinman (later Lidia Siciarz). Born in Poland in 1932, Lida was sent at the age of ten into hiding to live as a Catholic in orphanages, where she survived until the end of the war. The exhibit records her adult remembrance of that time: "We went to Warsaw...we were dispersed in different homes...Sister Sophia called me and said to me, 'There's a priest over there and you have to go and take communion...he is going to ask you...if you are Jewish....' She simply told me, 'you just have to lie and it's not going to be a big sin, God will understand.'"

Simon Jeruchim—who in 2001 published his memoir *Hidden in France: A Boy's Journey under the Nazi Occupation*—was one of the thousands of other Jewish children who passed for Christian to survive the war. He was born in Paris in 1929. During the Nazi occupation of France, Simon and his younger siblings were sent into hiding on farms in Normandy. He was reunited after the war with his brother and sister, and in 1949, they began new lives in the United States. However, they never saw their parents again, and would have to wait half a century before learning that they had been killed at Auschwitz.

In a recording available on the museum's website, Simon recalls working as a farmhand for a devoutly Christian family in Normandy:

> These people were very religious…They didn't ask any question whether I was a devout Christian or not, but it was taken for granted that I would be a Christian, so every night we had to pray…this was a hard floor, you get on your knees, and without long pants it is tough on your knees…Of course at the beginning I would mumble…Eventually I borrowed—I won't say I stole—I borrowed a prayer book when I was guarding the cows and just memorized everything in one sitting…and then I knew all the prayers even better than they did…We went to church once a week…I don't think she ever thought I was Jewish…they didn't even know what Jews would look like. They were so backwards…They didn't read newspapers. Those who had radios were the rare ones. Of course they knew there was such a thing as Jews but for them they bought into the myth that Jews had horns and since I didn't have horns or a tail, I was okay.

The suppression of one's blood or ancestral identity —
even if it is voluntary — can exact a serious price. Hahn
Beer writes about her painful efforts to reintegrate with
the Jewish community in the immediate aftermath of
the war. She met Jewish men who had been held in concentration camps during the war and spewed vitriol at
her upon learning that she had survived by marrying a
Nazi officer.

In other cultures and racial groups too, the act of
passing can create great pain, either for the ones who
have passed or for their children. The Canadian writer
Wayne Grady, author of numerous non-fiction books,
recently released his first novel, *Emancipation Day*,
which dramatizes the psychological toll exacted by passing. The protagonist, Jack Lewis, has grown up in a black
family in Windsor, Ontario, but when the novel opens he
is passing for white in Newfoundland during World War
II. Lewis falls in love with a local white woman, marries
her, and after the war moves with her back to Windsor —
without coming clean about his family background.

Emancipation Day draws its title from a day by the
same name in Canada — August 1, which is celebrated
by many African-Canadians as the anniversary of the
British abolition of slavery in 1834. Grady's novel hinges
on the tension created by Jack's secret, and by his active
suppression of the truth about his own family. Grady
wrote the novel years after discovering, in mid-life, that
his own father was black and had kept this secret from
his son.

"I was about fifty when I learned that my father's family (i.e. my family) were members of Windsor's black community, and that my father had passed for white during the war," Grady said to me in an email. "I was never able to talk to my father about my discovery: I tried, but he continued to deny any knowledge of his family's history. 'News to me,' is all he would say. How did I feel? Right from the beginning I was fascinated, overjoyed is not too strong a word here: I had grown up with no extended family, since my father's passing meant that he had turned his back on his parents and brothers and sister, and I had never met any of them except when I was too young to ask awkward questions. I felt I had suddenly been given a past. I was angry with my father for having denied me a family, a history, and I was further angry with him for continuing to refuse to tell me anything about his own past. I understand it — passing involves an enormous capacity for self-deception and denial — but it still frustrated and angered me. In fact, that anger got in the way of writing the novel for a good ten years. I had to get over my anger at my father and treat Jack Lewis as a character, with enough good qualities to explain why Vivian would fall in love with him in the first place, and stay with him when the truth began to come out."

One of the emotional problems related to passing is that, to succeed, you must eradicate your past and convince all people who matter of the "truth" of your new, fictional identity. In the case of surviving the Holocaust, escaping slavery, or avoiding the ravages and restricted possibilities associated with racial hatred and racial segregation, the act of passing is serious business. You have to succeed. If you are caught, the consequences may be severe. You must be on guard at all times against the possibility of being betrayed by your blood.

For example, Belle da Costa Greene—librarian to the American financier and art collector J. P. Morgan, and the first director of the Pierpoint Morgan Library—was born of African-American parents in Washington, D.C., in 1883. After her parents separated, Belle and her mother and siblings passed for white, changing their names. Her mother changed her maiden name to Van Vliet to pretend that she had Dutch ancestry, and Belle took on the middle name da Costa as a way to explain her looks, considered by some to be exotic. For a woman who developed a reputation as an effective and influential librarian, and who was committed to preserving records—not destroying them—Belle da Costa Greene took an unusual step shortly before she died in New York City in 1950: she burned her personal papers. Imagine how strongly one must want to incinerate one's secrets to expunge them not just from the living record but posthumously as well.

THERE ARE COUNTLESS STORIES about racial passing in North America. Anatole Broyard, the celebrated *New York Times* literary critic who died in 1990, was born of African-American parents in New Orleans and raised mostly in Brooklyn. He aspired to write but did not want to be defined by his blackness, so he took advantage of his light skin colour (both of his African-American parents were Louisiana Creoles) and passed for white. The journalist Brent Staples wrote in the *New York Times* in 2003 that Broyard wanted to be a writer — and not just a "Negro writer," consigned to the back of the literary bus. Writing for the *New Yorker,* the Harvard historian Henry Louis Gates noted that Broyard "did not want to write about black love, black passion, black suffering, black joy; he wanted to write about love and passion and suffering and joy." And so Broyard constructed a white identity for himself. He did not inform his own children that he was black. His daughter, Bliss Broyard, wrote in her memoir, *One Drop,* about how Broyard — even when he had advanced prostate cancer and his body was wracked with pain — resisted telling her brother Todd and her about his secret ancestry. *One Drop* opens with a story from two months before Broyard died, when he met with his wife, Alexandra, and their children, Bliss and Todd, in their home in Martha's Vineyard. When Alexandra asked if there was anything that the ailing Anatole would like to tell his children, Anatole prevaricated. When his wife persisted by announcing that Anatole had lived with a secret for a long time, Anatole tried to shut her up by saying, "Goddamn it, Sandy."

A man who spent his life spinning words on the page, and became one of the most highly respected literary critics in the United States, could not, even with death at the door, say the three words "I am black." When both children pressed him to tell the secret to which their mother alluded, Broyard said he would tell them, but not that day. "I need to think about how to present things," he said. "I want to order my vulnerabilities so they don't get magnified during the discussion." But he never did answer their questions.

The distance between ancestral and constructed identities haunted Broyard beyond the grave. Much earlier in life, he had married a black Puerto Rican woman named Aida Sanchez. Together they had a daughter named Gala. The couple divorced after Broyard returned from military service in World War II. Subsequently, Broyard married Alexandra Nelson, a white American woman of Norwegian heritage. They raised their children as white. The day after Broyard died, in 1990, an obituary ran in the *New York Times* — the same newspaper for which he had worked as book critic, essayist, and editor. The obituary contained 811 words, but it did not include a single mention of his first wife, of his first and eldest daughter, of their African ancestry, or of his. And, as Bliss Broyard informs readers in *One Drop*, Broyard's death certificate lists him as white.

Gates, the Harvard historian, argues that in shelving his true identity in order to inhabit one that he purposefully constructed, Broyard failed to live up to his greatest potential as a writer. He lost touch with himself, Gates argues, and thus he lost the ability to write profoundly. He had shown great promise as a creative writer, but ultimately settled into the life of reviewing books by other people.

Even by his own children, Broyard was taken to be white. For them, he lived and died as white. Did that make him white? Being of Creole ancestry, he had white ancestry as well as black. So who is to say that moving into one part of his identity — a part that would not normally be recognized, but that cannot factually be denied either — was fraudulent or wrong? The problem is that one could not be both black and white in America. Being in any way black made you black. For Broyard to become white, he could no longer acknowledge any blackness.

Philip Roth writes in his novel *The Human Stain* about the character Coleman Silk, a white man (from all appearances) who is hounded from his job as a university professor after inadvertently uttering a racial epithet in his classroom. It is an absurd situation, because Silk refers to some students who have never shown up in class as "spooks," by which he means ghosts. But as it turns out the missing students are black, and his unfortunate word choice is taken as a racial epithet. Over the course of the novel, we learn that Silk is an African-American who has passed as a white, Jewish man in order to decide his own fate in the world, rather than having others do that for him. Silk married a white woman but never told her or their four children about his racial background.

Some people have speculated publicly — including on *Wikipedia* — that the novelist Philip Roth may have drawn upon the life of Anatole Broyard to write *The*

Human Stain. Roth had met Broyard at least twice. Once, the two men met unexpectedly in a men's clothing store on Madison Avenue in New York City. Roth spontaneously bought him a pair of shoes as a playful mock ploy to curry the literary critic's favour. In an article published in 2012 in the *New Yorker*, Roth acknowledged that he had heard many years earlier that Broyard was secretly "an octoroon." (This term is meant to describe a person considered to have one-eighth black blood, although in reality Broyard's parents were both blacks.) However, the novelist vehemently refuted the suggestion that Broyard's life had inspired the creation of Coleman Silk.

Regardless of whether Roth's imagination was tinged by having known a fragment of the literary critic's personal background, the real Broyard and the fictional Silk certainly paid their dues for having passed. Neither man appears to have been truly in touch with himself. In the case of the real man, Broyard had every right to choose his path. He got the job with the *New York Times*, which he would have been unlikely to snag had the paper's editors known he was black. But he paid a price. He deprived himself of an open, intimate, publicly acknowledged connection to his own family heritage. In so doing, he robbed his children of the same opportunity, at least while he was alive and they were young and still shielded from the truth. Still more people lost out. He denied Americans of all races and backgrounds the chance to know his gifts and genius, in all their rich complexity. He buried a beautiful truth that might have galvanized dozens, hundreds, thousands or millions of people — but we will never know for sure, because Anatole Broyard died with his secret largely intact.

A famously unusual case of passing is found in the story of the Texan journalist and photographer John Howard Griffin, who decided in 1959 to attempt to document the real meaning of being black in America. Griffin, who was white, consulted a physician and underwent a series of treatments by means of drugs, sun lamps, and skin creams to make his skin look brown. He shaved his head so that his straight hair would not give him away. And he bused and hitchhiked into some of the most racist zones of the United States, such as New Orleans, Mississippi, South Carolina, and Georgia, to write about his experiences. His 1961 book, *Black Like Me*, became a bestseller and is still remembered a half-century later. Oddly, a white man passed for black to tell a primarily white audience what it truly meant to be black. Did John Howard Griffin truly become a black man? He was certainly considered one by the people he met in the course of his journalistic quest.

Most observers today would probably agree that Griffin was not black, but only posing as black. But what if, instead of choosing to undergo treatments to make his skin look brown as an adult, Griffin had been exposed to such treatments as an infant? What if his parents had found doctors willing to experiment on him? And what if he had then been orphaned and adopted as a black child, and his appearance had remained unchanged for the rest of his life? If he had lived out the entirety of his life as a black man, would he then have been legitimately black? I would have to say yes. He would have been legitimately black because the world would have seen him so. This has nothing to do with blood or biology, and everything to do with social interaction and the negotiation of one's identity in public and private spheres.

Although some may now consider Griffin's approach offensive, it can also be seen as a testament to the arbitrary nature of race, and to the insanity of imagining that racial identity is rooted in the nature of one's blood. John Howard Griffin had to hide his own white identity to investigate and later expose what it meant to be black in America.

Another reversal of racial identity took place in 1930 in Oakville, Ontario, when a young World War I veteran named Ira Johnson became engaged to a white woman named Isabella Jones. Johnson had grown up as black in Oakville and was widely considered black by the black community in town. His mother had been a midwife serving black mothers, and his family had attended a black church. However, when the Ku Klux Klan came into town burning crosses and threatening Johnson's life, he was naturally afraid. He invited the *Toronto Star* into his home and declared that he was not black, but of Cherokee descent. The *Toronto Star* bought it, entirely. On the front page of its March 5, 1930, edition, the *Star* ran an article with the headline "Is of Indian Descent Ira Johnson Insists: Oakville Man, Separated from His Sweetheart, Traces His Ancestry."

The article began with this lead paragraph: "Ira Junius Johnson, separated from his sweetheart, Alice Jones by Ku Klux Klansmen here last Friday, is of Indian descent and has not a drop of negro blood in his veins, he told the *Star* yesterday at the home of his mother, who is a refined and intelligent woman."

My own interviews in the 1990s with Alvin Duncan, a black resident of Oakville who was a teenager at the time of the KKK raid, confirmed that at the very least, the members of Oakville's black community had always assumed that Johnson was black. Perhaps he tried to establish a new, Cherokee identity to avoid the wrath of the KKK. He certainly would have had good reason to fear the group, which was well known for the lynching of blacks in the United States. On the other hand, there is the possibility that Johnson did have Aboriginal ancestry. Perhaps we will never know for sure, but the incident provides yet another example of the negotiation of racial identity and the lengths to which people will go to keep their blood ancestry secret because of persecution and pressure from the outside world. Tomorrow, perhaps, things will change. But today, race has nothing to do with blood, and everything to do with what people will believe.

FOR CENTURIES, RACE HAS COME to be equated with blood. But will modern science displace that notion? Over recent years, as the science of genetics has evolved, increasing numbers of people, hungry for details about their ancestral history, have begun having their DNA tested to unearth clues about their past. Science now holds out certain promises that seemed hitherto impossible.

DNA tests have shattered a myth that persevered despite all common knowledge to the contrary: that blacks were black and whites were white, and that a person could absolutely not be both. To have admitted such a thing, historically, would have been to do much more than to admit the awful truth that white slave masters took black slave women into their beds. It would have reduced to rubble the foundations of an economy and society based on the subjugation of one people by another. For how could one subjugate the other if they were truly the same?

One of the most famous examples, which has altered the way many historians have framed U.S. history, is that of the relationship between Thomas Jefferson and Sally Hemings. Hemings has long been described as the slave mistress and mother of many children fathered by Jefferson, who was the key author of the Declaration of Independence and the third American president, serving in office from 1801 to 1809. Historians, both professional and amateur, battled for centuries about the paternity of Hemings's children. Although some still debate the matter, it wasn't until the results of a DNA test were published in the magazine *Nature* in 1998 that historians and public institutions began to accept what many African-Americans had long believed: that those children were indeed fathered by Thomas Jefferson.

In 1802, the American journalist James Thomson Callender set off a political firestorm that has lasted more than two centuries when he published an article in the *Richmond Recorder* titled "The President Again."

It began with these words:

> It is well known that the man, *whom it delighteth the people to honor*, keeps, and for many years past has kept, as his concubine, one of his own slaves. Her name is SALLY. The name of her eldest son is TOM. His features are said to bear a striking although sable resemblance to those of the president himself. The boy is ten or twelve years of age. His mother went to France in the same vessel with Mr. Jefferson and his two daughters. The delicacy of this arrangement must strike every person of common sensibility. What a sublime pattern for an American ambassador to place before the eyes of two young ladies!

For having the temerity to suggest that the president of the United States had conducted a long-term affair with a black woman, Callender was vilified over the years as "obnoxious," "a liar," "a scoundrel," "tempestuous," "unsavory," and "generally odious" — to quote just a few epithets. The allegations did not appear to damage the political career of Jefferson, who was easily re-elected two years later.

Like many others, Jefferson's daughter Martha Jefferson Randolph denied that the president had had a relationship and children with Hemings. Two of Jefferson's grandchildren also claimed that, for moral and practical reasons, the liaison would not have been possible. In 1979, Jefferson historians persuaded the CBS television network to drop plans to air a television miniseries (based on the best-selling novel *Sally Hemings*, by Barbara Chase-Riboud) about the relationship. One 1979 letter to CBS chairman William Paley from Merrill Peterson — a prominent University of Virginia historian who had written a book dismissing the Jefferson-Hemings story, as created on "the flimsy basis of oral tradition, anecdote, and satire" — said the network should "reconsider lending its name and network to mass media exposure of what can only be vulgar sensationalism masquerading as history."

For his part, Thomas Jefferson was well known for taking strong public positions on slavery and miscegenation. He considered slavery an evil but owned hundreds of slaves himself, keeping many of them at his Monticello plantation in Virginia. In an outburst that spilled into his first draft of the Declaration of Independence, but which was later excised from the final version, Jefferson blamed Britain's King George III for the institution of slavery. The rebellious American colonies were waging their War of Independence against Britain when Jefferson wrote his first draft in 1776, so some of the vitriol can be chalked up to the fever of war. Still, Jefferson was adept at leaving himself and other American perpetrators of slavery out of the portrait when he wrote of King George III: "He has waged cruel war against human nature itself, violating its most sacred rights of life & liberty in the persons of a distant people who never offended him, captivating & carrying them into slavery in another hemisphere, or to incur miserable death in their transportation hither..."

On the page, Jefferson was also clear about the sin of miscegenation. In one of his most often repeated quotes on the subject, Jefferson wrote in 1814: "The amalgamation of whites with blacks produces a degradation to which no lover of his country, no lover of excellence in the human character, can innocently consent." Most historians agree that Jefferson was still involved with his slave mistress when he wrote those words. If it is true that Jefferson fathered Hemings's son Eston Hemings, the child would have been just six years old when Jefferson railed on the page about the evils of miscegenation.

The details of Sally Hemings's life are on one hand astounding, and on the other hand they profoundly reflect experiences of many women enslaved in the Americas. Dolley Madison, the wife of James Madison (the fourth American president, who took office in 1809, after Jefferson's presidency ended), is widely quoted as saying that the Southern white wife was "the chief slave of the master's harem."

To complicate the interracial nature of Jefferson's Monticello — a five-thousand-acre tobacco plantation where about 130 slaves worked at any one time — Sally Hemings was actually the half-sister of Jefferson's wife, Martha Wayles Skelton. Hemings's father was a white planter by the name of John Wayles, who had Sally (and five other children) by his long-time slave mistress, Betty Hemings, and Martha (wife of Thomas Jefferson) by his first wife, Martha Eppes. Sally herself was light-skinned and considered a "quadroon" — meaning that one of her four grandparents was black and the other three were white.

At the age of about fourteen, Hemings travelled to Paris to serve Jefferson for about two years in his capacity as the American ambassador to France. Her sexual relationship with him is thought to have begun there, or shortly after their return to the United States. According to the Thomas Jefferson Foundation, which maintains a museum on the grounds of Jefferson's former plantation, "Most historians believe that, years after his wife's death, Thomas Jefferson became the father of the six children of Sally Hemings mentioned in Jefferson's records, including Beverly, Harriet, Madison, and Eston Hemings."

Hemings's relationship with Jefferson is thought by many historians to have lasted for nearly forty years, until Jefferson's death in 1826. The man who entered meticulous notes about the fathers of his slaves did not record in his plantation slave register any name for the father of Hemings's children at Monticello. Although he did not free Hemings while he lived or in his will, he freed her children and other relatives — the only nuclear slave family on his massive estate to be freed — and asked the Virginia legislature to allow them to remain in the state after they were freed. (Normally, freed blacks had to move to another state.)

In this case, DNA was able to put to bed (more or less) a centuries-old debate about blood and ancestry. What was disputed in the blood was resolved (most believe) by means of genetic analysis. Perhaps the fact that the chapter is closing on this two-hundred-year-old debate signifies that we will be able to move away from thinking about racial identity as a function of blood.

GENETIC RESEARCH TODAY takes us far beyond historical debates about who fathered whom in the nineteenth century, and gives ordinary citizens the hope of pinpointing ancestral connections that they might not otherwise discover.

Many people long to know more about the secrets of their ancestry. When we have children, and as we age, we often crave more details about our distant family history. It's part of how we come to know and identify ourselves, and part of the legacy that we want to pass on to our children. For the centuries that genealogists have been digging into family histories, they have been largely confined to paths connected to blood ancestors and the people who came into those ancestors' lives. Who were our parents? Grandparents? Great-grandparents? How about our great-great-great-grandparents? Going seven generations back, we have 128 parental relations with five *great*s before the word *grandparent*. How many of us know some or all of them? How many of us are able to leap across continents and seas and trace distant ancestors in distant lands?

The blood trail dries up the further back we go, and many people — especially those whose lives or whose ancestors' lives have been interrupted by forced migration, genocide, slavery, or other such misfortune — lack the combination of dumb luck, contacts, tools, and skill to map our family tree many generations back. Regardless of your ancestry, if the knowledge of your family tree runs into a roadblock two, three, seven, or ten generations back, DNA testing offers the possibility of knowing what was previously unknowable, to behold the breadth of your history in ways that had once seemed unthinkable.

Many companies have sprung up to take saliva from clients, analyze the DNA, and not just tell them about the physical locations of their ancestors, but to name people with whom they may share ancestry. The Harvard historian Henry Louis Gates ran a television show called *African American Lives*, which offered to unearth — via genetic analysis — the family history of numerous famous African-Americans. His book *In Search of Our Roots* explores the same thing. Gates and a team of researchers and experts combined old-fashioned genealogical sleuthing with DNA tests on Oprah Winfrey, Whoopi Goldberg, Tina Turner, Maya Angelou, and other celebrities. In each case, Gates presented them with information about the "admixture" of their ancestral heritage. In addition to her African ancestry, Oprah Winfrey was declared to have 11

percent Native American ancestry. Analysis of her mitochondrial DNA (reflecting maternal lineage) revealed that
she had genetic traits in common with people in Liberia,
Cameroon, and Zambia, as well as the Gullah people of
South Carolina. Gates told Whoopi Goldberg that she
had an ancestral admixture that was 92 percent sub-
Saharan African and 8 percent European. The European
admixture in Goldberg's heritage was lower than usual
for African-Americans, who, Gates said, usually have
about 20 percent European heritage. In Goldberg's case,
DNA testing along her matrilineal line revealed that she
shared genetic signatures with people from Sierra Leone,
Liberia, and Guinea-Bissau. In concluding his chapter
on Goldberg, Gates writes in favour of exploring one's
genealogical background: "This sort of knowledge can
ground you... Knowledge of your ancestry can provide a
certain sense of calm about the past, where before there
were only questions — hundreds of years of unanswered
and seemingly unanswerable questions."

So far, there are limits to what genetic ancestry tests
can tell us. In her book *The Juggler's Children: A Journey
into Family, Legend and the Genes That Bind Us*, the *Globe
and Mail* science journalist Carolyn Abraham tells us
that when she began exploring her own family history,
the first genetic test she encountered declared that she
was of 22 percent Native American ancestry. When she
investigated this detail, she wrote, she was told that the
portion of her ancestry that had been presented to her as
Native may just as well have been Asian.

Troy Duster, a sociologist at New York University, warns that genetic testing—for family ancestry, or for the purposes of determining the racial profile of suspected criminals—offers up details that are taken as fact, but which are often based on partial information or derived from subjective or racially biased starting points. He has explored the issues in a book entitled *Backdoor to Eugenics* and an essay entitled "Ancestry Testing and DNA," published in 2011 in the anthology *Race and the Genetic Revolution: Science, Myth, and Culture*. In the essay, Duster notes that since 2002, nearly half a million people have purchased ancestry-tracing DNA testing kits from at least two dozen companies in the market. Henry Louis Gates was told by one company that his maternal lineage traced back to Egypt, but by another company that his maternal ancestors were neither Egyptian nor African, but most likely European. An African-American woman, he said, took three DNA tests and was given three different results: depending on the test, her ancestors were located in Sierra Leone, the Ivory Coast, or Senegal.

By analyzing the Y chromosome, Duster says, it is possible to determine whether a certain male descends from a specific father, grandfather, great-grandfather, and so on, along paternal lines. Through analysis of mitochondrial DNA, it is also possible to determine whether a woman descends from a specific mother, maternal grandmother, and other maternal ancestors. Mitochondrial DNA tests helped re-establish family connections in Argentina, after the dictatorship between 1976 and 1983 forcibly removed children from their parents, murdered the parents, and gave the children into the care of adoptive parents. When the "Dirty War" ended, DNA testing helped reunite grandmothers with their rightful grandchildren.

However, Duster says, the two genetic tests along sex-linked ancestral lines will point to only two among a vastly increasing number of distant ancestors. For example, looking back over eight generations, each person has 256 great-great-great-great-great-great-grandparents. Sex-linked ancestral tests for males and females could be used to identify only two out of those 256 ancestors. The other ancestors also contributed to any descendant's genetic makeup, but they are left out of the picture when we are told about our distant, genetically defined ancestors. Therefore, just because a sex-linked ancestral test does not suggest genetic matches with a certain group of people — say, Seminoles in the United States or the Mende people of Sierra Leone — it does not mean that the person being tested does not have ancestors from that group. It just means that a very limited test did not turn up a positive connection.

In addition, Duster casts doubt on the reliability of genetic tests that purport to establish "ancestry information markers," which claim to tell a person what *portion* of his or her ancestry is, say, sub-Saharan African, European, or East Asian. These tests have several problems, according to Duster: they rely on the dubious idea that certain groups of people have shared genotypes that are pure and distinct from those of other groups of people; they suggest some ancestral connections while ignoring other possibilities; and, although they are based on genetic tests comparing the subject's DNA to that of contemporary people living in distant places, the ancestors of those same contemporary people may have migrated over many centuries.

DNA analysis can be used to determine a person's direct paternal or maternal lineage, Duster concludes, but "claims to determine links to ancestral populations of many prior centuries must be necessarily incomplete, tentative, speculative, and of little use... Since we are witnessing a surge in ancestry testing across the globe, the best advice for the unsuspecting consumer is *caveat emptor* (buyer beware)."

Genetics has opened a door to further explanation of our ancestral identities, but it appears to be just as fraught and contested as blood. It is only natural that we should look to science, genealogical sleuthing, direct knowledge of our family ancestors, and all other possible means to put together the ever-shifting puzzle of our identities. Just as poetry and paintings help define us, we also look to family trees to enrich our notions of identity. Genetics, in the year 2013, has not displaced blood as a means to tell us who we are. It helps complicate the picture, and that will be a good thing if it forever shatters the notion that one person's blood may be "pure" Catholic (during the Inquisition), white (during the time of slavery), Aryan (during the Holocaust), or heterosexual (during tainted blood scandals that rocked Canada, the U.K., France, the United States, and Japan), and therefore superior to the blood of Jews, Muslims, blacks, gays, or others.

WE IMAGINE SCIENCE TO BE PURE, inviolable, and absolutely true, but we have only to look at the evolving theories of human blood and human circulation, over the millennia, to realize that scientists — like everyone else — move at least partly in step with the social biases and subjective limitations of their time.

Science will contribute to our understanding of who we are, and of how blood enters into the picture. Perhaps one day we will find a cure for AIDS, so that we no longer have to experience the dreadful fear that the stuff inside our veins will turn toxic and kill us if we receive a tainted transfusion. There will be other diseases down the road, however, and some of them are surely to be carried by blood, passed from human to human through transfusion, sexual intimacy, mosquitoes, or some other vector. Two hundred years ago, we didn't even know that mosquitoes transmitted malaria by sucking virus-laden blood from one human and biting another. Who can say what we will know in two hundred years?

Let's imagine that we are still roaming the planet two centuries from now, writing books, designing buildings, making love, combing the nits out of children's hair, dethroning corrupt politicians, paddling canoes, and being seduced by the Northern Lights. Let's say that we have not succumbed to complete and utter folly and exterminated ourselves through war, environmental destruction, or disease. If we do survive as a species for another two hundred years — enough time for our next six generations of descendants to be born and to procreate — we will surely look back with a certain amusement and astonishment at how humans were thinking about blood today.

Looking back, we will be no less amused and horrified than we are now when we remember that as recently as the eighteenth century, a French physician was trying to calm a man's psychiatric disorders by transfusing the blood of a calf into him. Just imagine how people may speak of us, in the future. *There was this thing called leukemia*, people might say. *They had the insane idea of blasting the body with toxins and damn near killing a person, and then replacing the bone marrow so that the body could manufacture an entirely new batch of blood. It was not until the year 2079 that a scientist by the name of Artemisia Peters of Zambia discovered that instead of x, y, z, all one had to do was a, b, c. And it was not until the year 2124 that the Chinese physician Ling Xiabo was awarded the Nobel Prize for discovering how to create safe, effective artificial blood that ended up eliminating the need for human blood transfusions.*

Who knows what we will know in two hundred years? None of us will be around to see it. But we are still likely to be thinking about our blood, not just as a function of our health and vitality, but also in regard to how it defines us in our families and countries, and in our personal and collective identities. Who do we descend from? Who do we belong to? What do we hope to transmit to our children and grandchildren, and to their offspring? How does blood come to be associated with truth and integrity? Will we ever transcend the nasty tendency to tumble into depravity by vilifying as impure the blood of the ones we most fear or revile? We have nobody but ourselves to blame for these lapses in humanity. There is nothing but our own biases blocking the way to a path that allows us to enjoy blood as a metaphor for our distinctiveness and group belonging, without using it as an excuse to pillory the most convenient scapegoat. Blood, I hope, will eventually unite us. Blood fills our imaginations just as fully as it fills our veins. Thus it has always been, and will always be.

ANNOTATED BIBLIOGRAPHY

CHAPTER 1

PAGE 13: "DISOBEDIENCE" BY A. A. MILNE

I have quoted the opening lines of A. A. Milne's poem "Disobedience," in *When We Were Very Young* (Methuen & Co., 1924).

PAGES 13–24: THE NATURE AND FUNCTIONS OF BLOOD

Sarah Levete, *Understanding the Human Body: Understanding the Heart, Lungs and Blood* (Rosen Publishing Group, 2010). It never hurts to start with a book for children.

Volume 1 of Justice Horace Krever's *Report of the Commission of Inquiry on the Blood System in Canada,* tabled in 1997.

Alistair Farley, Charles Hendry, and Ella McLafferty, "Blood Components," *Nursing Standard,* November 28 – December 4, 2012.

Jacques-Louis Binet, *Le sang et les hommes* (Gallimard, 2001). See the photos on the first pages, which show how quickly blood clots after a vein has been cut.

The PBS web page "Red Gold: The Epic Story of Blood" offers a wealth of physical and historical facts at www.pbs.org/wnet/redgold/index. html.

For notions of the four humours advanced by Hippocrates, Galen, and others, see: Noga Arikha, *Passions and Tempers: A History of the Humours* (Ecco, 2008), and Sherwin B. Nuland's review "Bad Medicine," *New York Times Book Review*, July 8, 2007.

PAGES 24–28: BLOODLETTING

Gerry Greenstone, "The History of Bloodletting," *B.C. Medical Journal*, January/February 2010.

Melissa Jackson, "The Humble Leech's Medical Magic," BBC News online, July 2, 2004.

PAGE 26: BLOODLETTING IN TALMUDIC TIMES, AND THE REFERENCE TO MAIMONIDES

Fred Rosner, "Bloodletting in Talmudic Times," *Journal of Urban Health: Bulletin of the New York Academy of Medicine* 62, no. 9 (November 1986).

PAGES 26–27: FAMOUS PEOPLE WHO DIED DURING BLOODLETTING

Liakat Ali Parapia, "History of Bloodletting by Phlebotomy," *British Journal of Haematology* 143, no. 4 (November 2008).

PAGES 28–29: WILLIAM HARVEY AND BLOOD CIRCULATION

The quote about William Harvey shaking up the seventeenth-century medical establishment comes from page xii of Thomas Wright, *Circulation: William Harvey's Revolutionary Idea* (Vintage Books, 2012).

PAGES 29–34: BLOOD TRANSFUSIONS

Just a few hundred years ago, we were transfusing animal blood into humans. Details about the seventeenth-century adventures in blood transfusions in Paris were drawn from: Douglas Starr, *Blood: An Epic History of Medicine and Commerce* (Quill, 2000) and Holly Tucker, *Blood Work: A Tale of Medicine and Murder in the Scientific Revolution* (W. W. Norton, 2011).

PAGES 32–33: NORMAN BETHUNE

Much has been written about the Canadian, who remains revered in China. For a summary of his contribution to blood transfusion advances in the Spanish Civil War, see: Peter H. Pinkerton, "Norman Bethune, Eccentric, Man of Principle, Man of Action, Surgeon, and His Contribution to Blood Transfusion in War," *Transfusion Medicine Reviews*, July 2007.

PAGES 34–45: MENSTRUATION

The observation about women's "defective barrels" comes from page 47 of Janice Delaney's book *The Curse: A Cultural History of Menstruation* (University of Illinois Press, 1988).

Since Delaney writes critically about how men in ancient times speculated about women's menstrual cycles, I found it helpful to check out the original comments. A.L. Peck translated Aristotle's *Generation of Animals* for Harvard University Press in 1943.

Martha K. McClintock, "Social Control of the Ovarian Cycle and the Function of Estrous Synchrony," *American Zoologist* 21, no. 1 (1981).

Mark A. Guterman, Payal Mehta, and Margaret S. Gibbs, "Menstrual Taboos Among Major Religions," *Internet Journal of World Health and Societal Politics* 5, no.2 (2008).

Allyn Gaestel, "Women in Nepal Suffer Monthly Ostracization," *New York Times* (online), June 14, 2013.

Tom Porteous, "'I Need Feminism Because...': in Pictures," *The Tab*, April 23, 2013, http://cambridge.tab.co.uk/2013/04/23/i-need-feminism-because-in-pictures/.

Richard Neill, Facebook post, October 8, 2012, www.facebook.com/Bodyform/posts/10151186887359324.

"Bodyform Responds: The Truth," www.youtube.com.

Arwa Mahdawi, "Bodyform's Bloodless Snark Attack," *Guardian* (Manchester) online, October 17, 2012 www.guardian.co.uk/commentisfree/2012/oct/17/bodyform-bloodless-snark-attack.

Stephanie Nolen, "India's Improbable Champion for Affordable Feminine Hygiene," *Globe and Mail*, October 3, 2012.

Gloria Steinem, "If Men Could Menstruate," in *Outrageous Acts and Everyday Rebellions* (Holt, Rinehart and Winston, 1983).

PAGES 50–51: IGNAZ SEMMELWEIS

For the life and struggles of Ignaz Semmelweis and his efforts to prevent blood poisoning in maternity wards in the early and mid-nineteenth century in Austria, I first drew upon a book by my maternal grandfather: George Bender, *Great Moments in Medicine* (Parke-Davis, 1961).

To be sure that my grandfather wasn't telling a tall tale, I kept looking:

Patrick Berche and Jean-Jacques Lefrère, "Ignaz Semmelweis," *La presse médicale* online, January 2011, www.em-consulte.com/revue/lpm.

S. W. B. Newsom, "Pioneers in Infection Control: Ignaz Philipp Semmelweis," *Journal of Hospital Infection* 23, no. 3 (March 1993).

PAGES 51–55: RH DISEASE (HEMOLYTIC DISEASE)

For details about early efforts to combat Rh disease (hemolytic disease), which remained the leading cause of neonatal and infant mortality until the middle of the twentieth century, I spoke with Raymonde Marius of Winnipeg, who donated plasma more than 1,000 times over the course of 40 years. I also spoke with Cheryl Lawson of Cangene Corporation in Winnipeg.

For details about the pioneering medical work of Dr. Bruce Chown of Winnipeg, who helped find a way to prevent the type A blood of a mother from attacking the type A+ blood of her fetus, see the article and video on the Canadian Medical Hall of Fame website: www.cdnmedhall.org/d-bruce-chown.

More details about Bruce Chown can be found in C. Peter W. Warren, *The Birth of a Medical Research Programme: The Rhesus (Rh) Factor Studies, Dr. Bruce Chown, and the Faculty of Medicine, University of Manitoba, 1883–1946*, an unpublished Ph.D. thesis for the Departments of History, Universities of Manitoba and Winnipeg, 2011.

Additional details came from the "Blood Components" article mentioned above in *Nursing Standard*, and from Kym H. Kilbourne's article "RHO(D) Immune Saves Thousands of Lives," *The Source*, Winter 2010.

CHAPTER 2

PAGES 67–69: PAULA FINDLAY

If you are interested in what can go wrong with the human body on the day of one of its biggest tests — a triathlon in the Olympic games — see this blog post: Paula Findlay, "A Series of Unfortunate Events," September 11, 2012, http://paulafindlay.blogspot.ca.

PAGES 76–80: HUMAN SACRIFICE

Entire books are devoted to human sacrifice, and I relied on some of them to write just a few pages on the subject. Among the most helpful were:

Mark Pizzato, *Theatres of Human Sacrifice: From Ancient Ritual to Screen Violence* (State University of New York Press, 2005). In his first chapter, Pizzatto describes the theatrical element of human sacrifice in ancient cultures.

In *Other Others: Levinas, Literature, Transcultural Studies* (SUNY Press, 2010), Steven Shankman offers an interesting meditation on how Caravaggio and Rembrandt offered contrasting paintings about the Biblical story of Abraham and Isaac.

Miranda Aldhouse Green, *Dying for the Gods: Human Sacrifice in Iron Age and Roman Empire* (Tempus Publishing Group, 2001).

For an article that explains the story of the kamikaze pilots during World War II and challenges the notion that all such pilots were happy to give up their lives for Japan's war effort, see David Powers, "Japan: No Surrender in World War Two," BBC History website, February 17, 2011, www.bbc.co.uk/history/worldwars/wwtwo/japan_no_surrender_01.shtml.

PAGES 80–82: HONOUR KILLINGS

For information about honour killings, see two articles, the first about the killings as an international phenomenon and the second about how they have been unfolding in Canada:

Robert Fisk, "The Crime Wave That Shames the World," *Independent* (London), September 7, 2010.

Marie-Pierre Robert, "Les crimes d'honneur ou le déshonneur du crime: étude des cas canadiens," *Canadian Criminal Law Review* 16 (2011).

PAGES 82–83: GREEK MYTHOLOGY

Marie Carrière, *Médée protéiforme* (University of Ottawa Press, 2012).

See *Wikipedia* article on Uranus at http://en.wikipedia.org/wiki/Uranus_(mythology).

See *Wikipedia* article on Aphrodite at http://en.wikipedia.org/wiki/Aphrodite.

PAGES 84–87: ARTEMISIA GENTILESCHI AND *JUDITH SLAYING HOLOFERNES*

My daughter Eve Freedman introduced me to the paintings by the Italian Artemisia Gentileschi about the story of Judith slaying Holofernes (in the Book of Judith). In an effort to catch up with a sixteen-year-old art history buff, I began to read up on the story.

For a quick overview of Gentileschi's life: see "Artemisia Gentileschi," *Encyclopedia of World Biography* (www.encyclopedia.com).

If women wanted to paint in Gentileschi's seventeenth-century Italy, they had to be married to painters: David Platzer, "Feminist Icon? David Platzer Salutes an Exhibition That Demonstrates the Greatness of Gentileschi — in Both Her Painting and Her Life," *Apollo*, June 2012.

For an art historian's book on Gentileschi, see Mary D. Garrard, *Artemisia Gentileschi* (Princeton University Press, 1991).

Mary D. Garrard and Gloria Steinem wrote a flyer to criticize a film for trivializing the rape of Gentileschi: www.h-net.org/~women/threads/disc-inaccurate.html.

For a news article about the film: Jonathan Jones, "Screen: The historians called it rape. The filmmaker called it romance. No wonder the feminists are up in arms," *Guardian* (Manchester), May 29, 1998.

For the suggestion that Artemisia may have painted *Judith Slaying Holofernes* in response to her rape by Tassi, see Rachel Spence, "Artemisia Gentileschi: Story of a Passion, Palazzo Reale, Milan," *FT.com* November 23, 2011.

PAGES 88–89: TRUTH AND BEAUTY

Ann Patchett's memoir *Truth and Beauty* (HarperCollins, 2004) is a story of friendship between two women, one of whom survives a childhood bout with cancer of the jaw, but embarks on a tragic path.

PAGES 89–98: STEM CELLS

Eric M. Meslin, Director, Indiana University Center for Bioethics and Associate Dean for Bioethics, Indiana University School of Medicine, provided me with information about embryonic stem cells and their controversy. As the former Executive Director of the U.S. National Bioethics Advisory Commission, Meslin was responsible for the Commission's publication *Ethical Issues in Human Stem Cell Research* (September 1999).

For introductory material about stem cells, see the National Institutes of Health website at http://stemcells.nih.gov.

For details about stem cell research in Canada, see the article "Stem Cell Research" by Patricia Bailey in *The Canadian Encyclopedia*, www.thecanadianencyclopedia.com/articles/stem-cell-research.

Advita Fund USA offers information about bone marrow transplants and their history on its website, www.advitausa.org/bone-marrow-transplant-understanding-the-term-and-procedure.

For information about the National Marrow Donor Program based in Minneapolis, Minnesota, see *Wikipedia*, "National Marrow Donor Program."

For information about haematopoiesis, which is the formation of blood cellular components, see *Wikipedia*, "Haematopoeisis."

Eliane Gluckman, Head of the Department of Bone Marrow Transplantation at the Hôpital Saint-Louis in Paris, has written extensively on issues of bone marrow transplants, as well as about the uses of cord blood. Her article "A Brief History of Haematopoietic Stem Cell Transplantation" appeared in *ESH-EBMT Handbook on Haematopoietic Stem Cell Transplantation*, published by the European School of Haematology in 2012. Gluckman's article, which can be found online, notes that after the Americans dropped atomic bombs on Japan and ended World War II, scientists began to look for ways to protect people from radiation, which led to stem cell research and bone marrow transplants.

In addition to winning the Nobel Prize, E. Donnall Thomas — born in Texas in 1920 — met his future wife in the course of a snowball fight. It must be an effective way to launch a romance, because he and Dorothy Martin remained married until he died, sixty years after their wedding. For an article about E. Donnall Thomas and his groundbreaking scientific work, which culminated in 1969 with a bone marrow transplant using a matched sibling donor for a patient, see Denise Gellene, "E. Donnall Thomas, Who Advanced Bone Marrow Transplants, Dies at 92," *New York Times*, October 21, 2012.

The history of stem cell research also owes much to Canadians. Ernest McCulloch and James Till proved the existence of stem cells in 1961 and defined their properties in 1963, and thus became known as the "fathers of stem cell research." See the article "James Till 1931–, Ernest McCulloch 1926–2011," on the website of the Canada Science and Technology Museum, at www.sciencetech.technomuses.ca.

For more about Ernest McCulloch's work on stem cells, see *Wikipedia*, "Ernest McCulloch."

For details about the work of the Nobel Prize–winning American biologist James Thomson, who isolated stem cells from human embryos in 1998 and then showed in 2007 that it is possible to transform skin cells into human embryos, see Gina Kolata, "Man Who Helped Start Stem Cell War May End It," *New York Times,* November 22, 2007; and *Wikipedia,* "James Thomson (cell biologist)."

"The Stem Cell Research Controversy" was posted January 5, 2011, on the website *Stem Cell History*: www.stemcellhistory.com.

The CBC news report "Stem Cells FAQs," posted October 12, 2010, can be found at www.cbc.ca/news/health/story/2009/01/07/f-stemcells.html.

Wikipedia, "Stem Cell Controversy."

Opposed to embryonic stem cell research: The Center for Bioethics and Human Dignity at Trinity International University, Deerfield, Illinois: http://cbhd.org/stem-cell-research/overview.

In favour: Francisco D. Lara, "Should We Sacrifice Embryos to Cure People?" *Human Affairs* 22, no. 4, October 2012.

"Shinya Yamanaka: Facts," on *Nobelprize.org,* http://www.nobelprize.org/nobel_prizes/medicine/laureates/2012/yamanaka-facts.html.

Matt Ridley, "Mind & Matter: Stem Cells Without the Controversy," *Wall Street Journal,* December 8, 2012.

Dennis Normille, "First Clinical Trial with Induced Pluripotent Stem Cells Grows Closer," *ScienceInsider,* June 26, 2013.

PAGES 98–99: THE DESIRE TO DONATE BLOOD

Richard M. Titmuss, *The Gift Relationship: From Human Blood to Social Policy* (Pantheon Books, 1971).

Eric M. Meslin, Patrick M. Rooney, and James G. Wolf, "Health Related Philanthropy: Towards Understanding the Relationship Between the Donation of the Body (and Its Parts) and Traditional Forms of Philanthropic Giving," *Nonprofit and Voluntary Sector Quarterly* (Supplement), 2008.

Gene Curtis, "Donors Overwhelm Blood Banks after 9/11 Attacks," *Tulsa World*, August 29, 2011.

PAGE 100: TUSKEGEE SYPHILIS STUDY

Susan M. Reverby, *Examining Tuskegee: The Infamous Syphilis Study and Its Legacy* (University of North Carolina Press, 2009).

E. M. Meslin and D. Mathieu, "The Tuskegee Syphilis Study," in Robert H. Blank and Janna Merrick, eds., *Encyclopedia of US Biomedical Policy* (Greenwood Publishing, 1996).

Wikipedia, "Tuskegee Syphilis Experiment."

PAGES 100–106: CHARLES DREW AND BERNARD LOWN

During World War II, the American Red Cross first banned blacks from donating blood, and then, after facing an outcry from the black community, ruled that African-Americans could donate blood that would be segregated and thus not transfused into whites, as described in Spencie Love, *One Blood: The Death and Resurrection of Charles R. Drew* (University of North Carolina Press, 1996). I drew quotes from pages 155–58 about Drew's objections to the blood segregation policies. In addition to biographical material, Love's book explores myth-making in the United States, and how Americans came to believe falsely that Drew died of blood loss in a car accident after a hospital in North Carolina refused to treat him because he was black.

Although policies of racial segregation affected nearly every walk of life in the American South in the early and mid-twentieth century, Drew died after a serious car accident, despite being treated promptly and thoroughly at the hospital.

Another biography of Charles Drew was published eight years before Love's book: Charles E. Wynes, *Charles Richard Drew: The Man and the Myth* (University of Illinois Press, 1988).

See *Wikipedia*, "Charles R. Drew," http://en.wikipedia.org/wiki/Charles_R._Drew.

In an interview on the *British Medical Journal* website, Dr. Bernard Lown speaks with Elizabeth Loder about the end of segregation by race of blood donations in the Johns Hopkins Hospital, at www.bmj.com/multimedia/video/2012/05/28/bernard-lown-part-3-segregation-blood-bank.

Lown delves deeper into the issue on his blog (http://bernardlown.wordpress.com), in "Black Blood Must Not Contaminate White Folks" (Essay 25), September 3, 2011.

PAGES 109–14: BLOOD DONATION POLICIES REGARDING MEN WHO HAVE SEX WITH MEN

Who is and who is not allowed to donate blood? Answers to this question touch a raw nerve, because we tend to feel that they rank our value as human beings. Are we or are we not worthy of giving blood? Here are some sources of information:

The World Health Organization says that blood donations should not be accepted from men who have sex with men. See pages 87 and 88 of *Blood Donor Selection: Guidelines on Assessing Donor Suitability for Blood Donation* (2012), at www.who.int/bloodsafety/publications.

American Red Cross blood donation exclusion policies, including a lifetime ban on donations from men who have had sex with men: www.redcrossblood.org/donating-blood/eligibility-requirements.

Tara Sun Vanacore and Abigail Barnes, "Tainted: Why Gay Men Still Can't Donate Blood," *Atlantic*, October 1, 2012.

Wikipedia, "Gay Male Blood Donor Controversy."

Mark A. Wainberg, Talia Shuldiner, Karine Dahl, and Norbert Gilmore, "Reconsidering the Lifetime Deferral of Blood Donation by Men Who Have Sex with Men," *Canadian Medical Association Journal*, September 7, 2010.

Charlene Galarneau, "Blood Donation, Deferral, and Discrimination: FDA Donor Deferral Policy for Men Who Have Sex with Men," *American Journal of Bioethics* 10, no. 2 (February 2010).

Charlene Galarneau, "Review of Anne-Maree Farrell, *The Politics of Blood: Ethics, Innovation and the Regulation of Risk*," *American Journal of Bioethics* 13, no. 4 (April 2013).

For information about how tainted blood products from an Arkansas prison made it into the blood supply in Canada and other countries in the 1980s, contributing to tainted blood scandals on an international scale, see the documentary *Factor 8: The Arkansas Prison Blood Scandal*, written, directed, and produced by Kelly Duda (Concrete Films, 2006).

Sophia Chase, "The Bloody Truth: Examining America's Blood Industry and Its Tort Liability through the Arkansas Prison Plasma Scandal," *William and Mary Business Law Review* 3, no. 2 (August 2012).

Justice Horace Krever, *Report of the Commission of Inquiry on the Blood System in Canada*, tabled in 1997.

André Picard, *The Gift of Death: Confronting Canada's Tainted-Blood Tragedy* (HarperCollins Canada, 1995).

Steffanie A. Strathdee, Michael V. O'Shaughnessy, and Martin T. Schechter, "HIV in the Blood Supply: Nothing to Fear but Fear Itself," *Canadian Medical Association Journal*, August 15, 1997.

Ruling by Judge Catherine Aitken, Ontario Superior Court of Justice, *Canadian Blood Services v. Freeman*, September 8, 2010, Court file 02-CV-20980.

Durhane Wong-Rieger, president, Canadian Association for Rare Disorders, interviewed on *Ontario Today* (CBC Radio), May 24, 2013.

Mexico lifted the ban on blood donations from men who have sex with men in 2012: www.care2.com/causes/mexico-no-longer-bans-gay-men-from-donating-blood.html.

Cheryl Wetzstein, "Study Could End Ban on Gay Men Donating Blood," *Washington Times*, May 16, 2012.

"U.K. to Lift Lifetime Ban on Gay Blood Donors," *Advocate.com*, September 8, 2011.

Peter Tatchell, the Australian-born U.K. activist on behalf of lesbian, gay, bisexual, and transgender people, argues on his website that the U.K. should accept blood donations from gay and bisexual males, provided that they always use a condom and test negative for HIV/AIDS: www.petertatchell.net/lgbt_rights/blood_ban.

This *Huffington Post* web page links to a CBC website and lists blood donation policies in twenty-one countries; Italy, Spain, and Mexico allow donations from men who have sex with men: www.huffingtonpost.ca/2013/01/12/gay-men-donating-blood_n_2467103.html.

Gillian Mohney, "FDA Ban on Gay Men as Blood Donors Opposed by American Medical Assocation," *ABC News*, June 20, 2013. See: http://abcnews.go.com/Health/american-medical-association-opposes-fda-ban-gay-men/story?id=19436366#.UefGcWT72Fd.

Art Caplan, "Opinion: Ease US blood supply shortage by lifting gay donor ban," *NBC News*, July 17, 2013. See: http://www.nbcnews.com/health/opinion-ease-us-blood-supply-shortage-lifting-gay-donor-ban-6C10656162.

PAGES 115–31: LANCE ARMSTRONG, BEN JOHNSON, AND PER-
FORMANCE-ENHANCING DRUGS

For questions and answers about blood doping, including details
on the potentially serious side effects, see the World Anti-Doping
Agency website, at www.wada-ama.org/en/Science-Medicine/
Science-topics.

Trent Stellingwerff, an exercise physiologist who works with the
Canadian Sport Institute, answered my questions about the effects
of exercise on bloodstream. He explained the ways that both legal
techniques (such as training at altitude or in hot weather) and illicit
approaches (such as blood doping) affect the blood and performance.
Stellingwerff assists in the coaching of his wife, Hilary Stellingwerff,
who is an international-calibre 1,500-metre runner. He also is part
of a team advising elite Canadian marathoner Reid Coolsaet. See the
website of the Canadian Sport Institute, at www.csipacific.ca.

Ex-professional cyclist Tyler Hamilton, a former teammate of Lance
Armstrong in the Tour de France, offers a lively, informative look at
blood doping and other forms of cheating in professional bike racing.
Hamilton openly describes his own use of drugs and blood doping
while riding with Armstrong. I drew quotes from pages 123 and 124 of
Hamilton's book: Tyler Hamilton and Daniel Coyle, *The Secret Race:
Inside the Hidden World of the Tour de France: Doping, Cover-Ups, and
Winning at All Costs* (Bantam Books, 2012).

In June 2013, shortly before the start of the 2013 Tour de France
(the first since Armstrong's admission five months earlier on
television to Oprah Winfrey that he had taken steroids and EPO and
undergone blood transfusions to cheat during his seven consecutive
victories), Armstrong told *Le Monde* newspaper that during the time
of his reign, it would have been impossible to win the race without
resorting to drugs and doping. See Stéphane Mandard, "Lance
Armstrong: Le Tour de France? Impossible de gagner sans dopage," *Le
Monde*, June 28, 2013, and Giles Mole, ed., "It Was Impossible to Win
Tour Without Taking Drugs, Claims Lance Armstrong," *Telegraph*
(London), June 28, 2013.

For a detailed presentation of Armstrong's cheating methods, as well as affidavits by Armstrong's former teammates, see the "Reasoned Decision" report by the United States Anti-Doping Agency: *Report on Proceedings under the World Anti-Doping Code and the USADA Protocol: Reasoned Decision of the United States Anti-Doping Agency on Disqualification and Ineligibility: United States Anti-Doping Agency, Claimant, v. Lance Armstrong, Respondent,* October 10, 2012, online at http://d3epuodzu3wuis.cloudfront.net/ReasonedDecision.pdf.

See the report of the formal inquiry struck after Canadian sprinter Ben Johnson tested positive for anabolic steroids at the 1988 Seoul Olympics and was stripped of his 100-metre world record and Olympic gold medal: Charles L. Dubin, Commissioner, *Commission of Inquiry into the Use of Drugs and Banned Practices Intended to Increase Athletic Performance* (Canadian Government Publishing Centre, Supply and Services Canada, 1990).

For details about events leading up to, during, and after Ben Johnson's 100-metre race at the Seoul Olympics, see Richard Moore, *The Dirtiest Race in History: Ben Johnson, Carl Lewis and the 1988 Olympic 100m Final* (Bloomsbury, 2012).

About the use of biological passports to detect blood doping or the use of performance enhancing drugs, see "Athlete Biological Passport," on the World Anti-Doping Agency website (www.wada-ama.org/en/Science-Medicine/Athlete-Biological-Passport); Mario Thevis, *Mass Spectrometry in Sports Drug Testing: Characterization of Prohibited Substances and Doping Control Analytical Assays* (Wiley, 2010); and Mario Zorzoli and Francesca Rossi, "Implementation of the Biological Passport: The Experience of the International Cycling Union," *Drug Testing and Analysis* 2, no. 11–12, 2010.

CHAPTER 3

PAGES 156–57: ANDERSON RUFFIN ABBOTT AND CATHERINE SLANEY

My father, Daniel G. Hill III, wrote about Canada's first black doctor, Anderson Ruffin Abbott (1837–1913), and in so doing introduced Catherine Slaney of Brampton, Ontario, to the fact that some of her ancestors had passed for white. Slaney had known that Abbott was an ancestor and a physician, but nobody had told her that he was black. See Catherine Slaney, *Family Secrets: Crossing the Colour Line* (Natural Heritage Books, 2003).

PAGES 159–64: INTERNATIONAL AND TRANSRACIAL ADOPTION

See Article 21 of the United Nations Convention on the Rights of the Child at www.ohchr.org/en/professionalinterest/pages/crc.aspx.

Wikipedia, "International Adoption" and "Transracial Adoption."

The National Association of Black Social Workers argues in a position paper against placing black children in non-black adopted families: www.nabsw.org/mserver/PreservingFamilies.aspx.

For a few details on Judge Edwin Kimelman's report on Indian and Métis adoptions and placements, which condemned as "cultural genocide" the widespread practice of removing Aboriginal children from their families and communities in Manitoba and sending them into adoption in eastern Canada or the USA, see *Wikipedia,* "Kimelman Report."

A *Wikipedia* article dealing with the "Sixties Scoop" and Judge Edwin Kimelman: "Sixties Scoop."

Adoption.com has an article on transracial adoption at http://encyclopedia.adoption.com/entry/transracial-adoption/360/1.html.

The Canadian Paediatric Society comments on transracial adoptions at www.cps.ca/documents/position/adoption-transracial.

The Evan B. Donaldson Adoption Institute weighs in on transracial adoptions at www.adoptioninstitute.org/publications/ MEPApaper20080527.pdf.

PAGES 165–67: BLOOD BROTHERS

Various *Wikipedia* articles offer details about the Norwegian warrior Örvar Odd and his Swedish counterpart Hjalmar: on the warriors, with the painting of them by Mårten Eskil Winge, see "Blood Brother"; on the story of Örvar Odd, with the painting by August Malmström, see "Örvar Odd."

PAGES 167–76: CITIZENSHIP

Gloria Galloway, "Autistic Girl's Future Up in the Air as Family Set to Be Deported from U.S., Refused Entry to Canada," *Globe and Mail*, December 28, 2012.

The quote about the notion of citizenship advanced by Romulus in Rome is taken from paragraph 16 of "The Life of Romulus," in Plutarch's *The Parallel Lives*, online at http://penelope.uchicago.edu/ Thayer/E/Roman/Texts/Plutarch/Lives.

For notions of *jus sanguinis* and *jus soli*, see Patrick Weil, "Access to Citizenship: A Comparison of Twenty-Five Nationality Laws," in *Citizenship Today: Global Perspectives and Practices*, ed. T. Alexander Aleinikoff and Douglas Klusmeyer (Carnegie Endowment for International Peace, 2001).

For the court case *United States v. Kim Wong Ark*, see *Wikipedia*, "*United States v. Wong Kim Ark*."

For the Chinese Exclusion Act in the United States, see *Wikipedia*, "Chinese Exclusion Act."

For an article about U.S. citizenship laws, see *Wikipedia*, "United States Nationality Law."

Audrey Macklin, "Who Is the Citizen's Other? Considering the Heft of Citizenship," *Theoretical Inquiries in Law* (May 2007).

Ann Gomer Sunahara, *The Politics of Racism: The Uprooting of Japanese Canadians During the Second World War* (Ann Gomer Sunahara, 2000), http://japanesecanadianhistory.ca/Politics_of_Racism.pdf.

PAGES 176–79: JAMA WARSAME AND SAEED JAMA

A report on Jama Warsame by the United Nations: *Jama Warsame v. Canada*, Communication No. 1959/2010, U.N. Doc. ccpr/C/102/D/1959/2010 (2011), International Covenant on Civil and Political Rights Human Rights Committee, 102nd session, July 11–29, 2011.

Audrey Macklin and Renu Mandhane, "Canada's New Exiles," *Ottawa Citizen*, November 25, 2012.

PAGES 181–82: GERMAN ANTHROPOLOGIST AND PHYSICIAN JOHANN FRIEDRICH BLUMENBACH

Johann Friedrich Blumenbach, *On the Natural Variety of Mankind*, www.blumenbach.info/_/Intro_to_Blumenbachs_Dissertation.html.

Lawrence Hill, *Black Berry, Sweet Juice: On Being Black and White in Canada* (HarperCollins Canada, 2001). See page 206 for the quote from Blumenbach.

Nell Irvin Painter, "Why White People Are Called 'Caucasian,'" Collective Degradation: Slavery and the Construction of Race, conference at Yale University, November 7–8, 2003.

Nell Irvin Painter, *The History of White People* (W. W. Norton, 2010).

Bruce Baum, *The Rise and Fall of the Caucasian Race: A Political History of Racial Identity* (New York University Press, 2006).

PAGES 182–87: THE *KOMAGATA MARU* STEAMSHIP AND THE STORY OF HARRY NARINE-SINGH

I drew heavily on the work of James Walker, an historian at the University of Waterloo, in writing *The Book of Negroes*. Once again, I have relied on his work to research this book, especially on matters of citizenship. See James W. St. G. Walker, *"Race," Rights and the Law in the Supreme Court of Canada* (Osgoode Society for Canadian Legal History and Wilfrid Laurier University Press, 1997). See especially chapter five, pages 246–95, *"Narine-Singh v. Attorney General of Canada."* The quote from McPhillips comes from page 261.

More information about Judge McPhillips's quote and decision can be seen in paragraph 102 of his decision on July 16, 1914, for the British Columbia Court of Appeal: *Canada v. Singh, Re Munshi Singh,* [1914] B.C.J. No. 116, 6 W.W.R. 1347, 20 B.C.R. 243.

For another analysis of the *Komagata Maru* steamship case, including a quote from Reverend Samuel Chown, see Audrey Macklin, "Historicizing Narratives of Arrival: The Other Indian Other," in *Storied Communities: Narratives of Contact and Arrival in Constituting Political Community,* ed. Hester Lessard, Rebecca Johnson and Jeremy Webber (University of British Columbia Press, 2010).

For details of the life of Reverend Chown, see an article by Neil Semple in *The Canadian Encyclopedia,* www.thecanadianencyclopedia. com/articles/samuel-dwight-chown.

For more about Reverend Chown and his quote, see Mariana Valverde, *The Age of Light, Soap, and Water: Moral Reform in English Canada, 1885–1925* (University of Toronto Press, 2008), 106.

In the course of her work in 1953–54 for the Toronto Labour Committee for Human Rights, my mother, Donna Hill, was involved in the early stages of Harry Narine-Singh's legal case. She answered my questions in a June 2013 interview.

PAGES 188–91: *CASTA* PAINTINGS AND PEOPLE OF MIXED RACE IN EIGHTEENTH-CENTURY MEXICO

The quote about mestizos comes from Octavio Paz, *Sor Juana; or, the Traps of Faith,* trans. Margaret Sayers Peden (Harvard University Press, 1988), 32.

Details about *casta* paintings and their historical context come from Ilona Katzew, *Casta Painting: Images of Race in Eighteenth-Century Mexico* (Yale University Press, 2004). The quote from José Gumilla is drawn from pages 48–49.

PAGES 191–95: MIXED-RACE ISSUES

Latin American mixed-race definitions come from Thomas M. Stephens, *Dictionary of Latin American Racial and Ethnic Terminology,* 2nd ed. (University Press of Florida, 1999), 71–72, 751.

The quote from Gunnar Myrdal comes from page 113 of Gunnar Myrdal, *An American Dilemma: The Negro Problem and Modern Democracy,* vol. 1 (Harper and Brothers, 1944).

Lawrence Hill, *Any Known Blood* (HarperCollins Canada, 1997).

See information about Strom Thurmond in *Wikipedia,* "Strom Thurmond."

Thurmond's quote insisting on the necessity of racial segregation can be see in the above *Wikipedia* article, as well as in Adrienne M. Duke, *Opaque Visions of the Self: The Possible Selves of African American Adolescent Males in the Context of Schooling* (ProQuest, UMI Dissertation Publishing, 2011), 20.

PAGES 195–201: IDENTITY ISSUES RELATING TO ABORIGINALS
AND BLACKS

Glen Coulthard, "Subjects of Empire: Indigenous Peoples and the
'Politics of Recognition' in Canada," *Contemporary Political Theory* 6,
no. 4 (2007).

Renisa Mawani, *Colonial Proximities: Cross-Racial Encounters and
Juridical Truths in British Columbia, 1871–1921* (University of British
Columbia Press/University of Washington Press, 2009).

Details about Virginia's Racial Integrity Act are taken from Kevin
N. Maillard "The Pocahontas Exception: American Indians and
Exceptionalism in Virginia's Racial Integrity Act of 1924." March 23,
2006/bepress Legal Series, Working Paper 1187, http://law.bepress.
com/expresso/eps/1187.

Troy Duster, "Lessons from History: Why Race and Ethnicity Have
Played a Major Role in Biomedical Research," *Journal of Law, Medicine
and Ethics*, Fall 2006. The quote is from page 494.

Circe Sturm, *Race, Culture, and Identity in the Cherokee Nation of
Oklahoma* (University of California Press, 2002). I have taken quotes
and details from pages 87–88.

Daniel Heath Justice, "Rhetorics of Recognition," *Kenyon Review*,
Winter 2010. I have taken quotes from pages 237 and 253.

Daniel Heath Justice, *Our Fire Survives the Storm: A Cherokee Literary
History* (University of Minnesota Press, 2006).

Details about mixed-race entries in the Canadian census in the
early twentieth century come from Chris Andersen, "From Nation
to Population: The Racialisation of 'Métis' in the Canadian Census,"
Nations and Nationalism 14, no. 2 (2008). Andersen also shared a
draft of his forthcoming book, *"Métis": Canada's Misrecognition of an
Indigenous People* (University of British Columbia Press, 2014).

Jean Teillet, "The Internal Migrations of the Métis of the Canadian Northwest," *Canadian Diversity,* Spring 2011.

PAGES 201–3: QUOTES FROM LOUIS RIEL

Jean Teillet, *Métis Law in Canada* (Pape Salter Teillet, 2012), 6.

Hold High Your Heads, the 1982 English translation by Elizabeth Maquet of *l'Histoire de la nation métisse dans l'Ouest,* by A. H. de Tremaudan (Pemmican Publications, 1936), 200.

Louis Riel, "Les Métis du Nord-Ouest [Regina]," in *The Collected Writings of Louis Riel/Les écrits complets de Louis Riel,* vol. 3 (University of Alberta Press, 1985), 278–79.

PAGES 203–6: STEVEN AND RODDY POWLEY IN THE SUPREME COURT OF CANADA

Various interviews by email and phone in 2013 with Jean Teillet (a Métis lawyer and partner in Pape Salter Teillet, www.pstlaw. ca/jeanbio.htm) and with Chris Andersen (a Métis scholar at the University of Alberta).

R. v. Powley, September 19, 2003, ruling by the Supreme Court of Canada, Citation 2003 SCC 43, File #28533.

Jean Teillet, *Métis Law in Canada* (cited above).

Factum of the Ministry of the Attorney General of Ontario, submitted in 1999 to the Ontario Superior Court of Justice in the case *Her Majesty the Queen v. Steve Powley and Roddy Charles Powley,* Court File 5799/99. I have drawn quotes and information from this factum, relying especially on paragraphs 21, 71–74, 77, 78, and 85.

PAGES 207–11: DEFINITIONS AND THOUGHTS ABOUT THE
RACIAL IDENTITIES OF INDIANS AND INUIT

Constance Backhouse, "The Historical Construction of Racial Identity
and Implications for Reconciliation," paper commissioned by the
Department of Canadian Heritage for the Ethnocultural, Racial,
Religious, and Linguistic Diversity and Identity Seminar, Halifax,
Nova Scotia, November 1–2, 2001.

Constance Backhouse, *Colour-Coded: A Legal History of Racism in
Canada, 1900–1950* (published for the Osgoode Society for Canadian
Legal History by University of Toronto Press, 1999).

Pamela D. Palmater is a Mi'kmaq lawyer whose family comes
from the Eel River Bar First Nation in northern New Brunswick.
She is an associate professor in the Department of Politics and
Public Administration at Ryerson University and Chair of Ryerson
University's Centre for Indigenous Governance. See www.ryerson.
ca/politics/facultyandstaff/bio_PamelaPalmater.htm and www.
nonstatusindian.com/bio/default.htm.

Pamela D. Palmater, *Beyond Blood: Rethinking Indigenous Identity*
(Purich, 2011). I have drawn quotes and details from pages 19, 28,
31–32, and 145.

PAGES 211–12: TRACEY DEER'S DOCUMENTARY FOR THE
NATIONAL FILM BOARD OF CANADA

Club Native: How Thick Is Your Blood? written and directed by Tracey
Deer (National Film Board of Canada, 2008). I have quoted from this
documentary about challenges to the status of four women living on
the Kahnawake First Nations reserve, which is located near Montreal.
See the reference to *Club Native* on the National Film Board website,
http://onf-nfb.gc.ca. Also see *Wikipedia*, "Tracey Deer."

PAGES 212–16: JUDGE MICHAEL PHELAN'S DECISION ON MÉTIS
IDENTITY FOR THE FEDERAL COURT OF CANADA

Ruling by Judge Michael J. Phelan, Federal Court of Canada, January
8, 2013, Docket T-2172-99, Citation 2013 FC 6, in the case known as
*Harry Daniels, Gabriel Daniels, Leah Gardner, Terry Joudrey and the
Congress of Aboriginal Peoples v. Her Majesty the Queen.* I have quoted
from paragraph 119.

Jean Teillet and Jason Madden, "Plainspeak on the Daniels Case
(Updated Version—February 2013)" (Pape Salter Teillet, 2013).

John Ibbitson, "Court Ruling on Aboriginal Peoples Opens a Pandora's
Box," *Globe and Mail,* January 9, 2013.

I have quoted Red Bear, who spoke in Toronto on the CBC Radio
program *Metro Morning* on June 10, 2013: http://www.cbc.ca/
metromorning/.

CHAPTER 4

PAGES 221–29: WITCHES

See the painting *Lilitu,* by Tara McPherson, at www.taramcpherson.
com/art/Paintings/Gallery%201/Detail/AD9F0B/Lilitu.

Jeffrey B. Russell and Brooks Alexander, *A New History of Witchcraft:
Sorcerers, Heretics and Pagans* (Thames and Hudson, 2007). I have
drawn especially from pages 4, 5, and 29.

Malcolm Gaskill, *Witchcraft: A Very Short Introduction* (Oxford
University Press, 2010). I learned much about the mistreatment of
witches in this pithy, accessible book by Gaskill, a professor of early
modern history at the University of East Anglia.

Gaskill's bio at the University of East Anglia: www.uea.ac.uk/history/
People/Academic/Malcolm+Gaskill.

Gaskill's website: http://malcolmgaskill.info/index.html.

Laura Stokes, "Prelude: Early Witch-Hunting in Germany and Switzerland," *Magic, Ritual, and Witchcraft* 4, no. 1 (Summer 2009).

Brian P. Levack, ed., *The Witchcraft Sourcebook* (Routledge, 2004).

Candace Savage, *Witch: The Wild Ride from Wicked to Wicca* (Greystone Books, 2000).

L. Frank Baum, *The Wonderful Wizard of Oz* (George M. Hill, 1900).

PAGES 229–33: FORCIBLE BLOOD REMOVAL

Scott Carney, *The Red Market: On the Trail of the World's Organ Brokers, Bone Thieves, Blood Farmers and Child Traffickers* (William Morrow, 2011). I have quoted from page 155.

The quote about the National DNA Data Bank is taken from the RCMP website, www.rcmp-grc.gc.ca.

The quote from the Brighteyes case comes from Jennifer Ditchburn, "Taking Suspect's Blood Violates Charter, Trial Told," *Vancouver Sun*, March 14, 1997.

For a U.S. Supreme Court ruling that limits but does not prohibit law enforcement authorities from ordering blood to be taken from people suspected of impaired driving, see *Missouri, Petitioner v. Tyler G. McNeely*, decision delivered April 17 by Supreme Court of the United States, 569 U.S. 2013, no. 11–1425 (2013).

PAGES 234–35: QUOTE FROM NIETZSCHE

Friedrich Nietzsche, *Thus Spoke Zarathustra* (originally published in German by Ernst Schmeitzner, c. 1883–85).

PAGES 235–41: SOR JUANA

Several people offered interviews and guided me to details about
the life of Sor Juana: Michael Schuessler; the Montreal poet and
translator Émile Martel; the Mexican poet and translator Pura López-
Colomé; and Pierre Sved and Shauna Hemingway of the Canadian
embassy in Mexico City.

Octavio Paz, *Sor Juana; or, the Traps of Faith,* trans. Margaret Sayers
Peden (Harvard University Press, 1988). I have drawn a quote
from Paz's book, which remains one of the most widely respected
biographies of Sor Juana.

Electa Arenal and Amanda Powell, *Sor Juana Inés de la Cruz: The
Answer/La Respuesta,* 2nd critical ed. and translation (Feminist Press
at City University of New York, 2009). This excellent book contains
biographical materials and translations of Sor Juana's memoir and
poetry. I have borrowed material and quotes from pages 15, 47, and
159.

Michael Schuessler, "The Reply to Sor Philothea," in *Latin-American
Literature and Its Times* (Moss, 1999).

Theresa Ann Yugar, *Sor Juana Inés de la Cruz: Feminist Reconstruction
of Biography and Text,* Ph.D. dissertation, Claremont Graduate
University, December 2012.

PAGES 241–42: DUELLING

"The First Duel Fought in Hot Air Balloons—Paris, 1808," *British
Newspaper Archive* blog, August 24, 2012; first reported on July 23,
1808, in the *Northampton Mercury*.

PAGES 245–48: BOXING

Stephen Brunt, *Facing Ali: The Opposition Weighs In* (Knopf Canada,
2002)

Bryan Weismiller and Tony Seskus, "Knockout, Manslaughter Trial Deliver Sucker Punch" *Calgary Herald*, June 9, 2012.

PAGE 250: BLOODHOUNDS

The information about Robert Boyle is drawn from *Wikipedia*, "Bloodhound."

PAGE 251–52: *IN COLD BLOOD*

Truman Capote, *In Cold Blood: A True Account of a Multiple Murder and Its Consequences* (Random House, 1965).

PAGES 252–53: *THE HUNGER GAMES*

Suzanne Collins, *The Hunger Games* (Scholastic Press, 2008).

PAGES 253–54: THE NATIONAL RIFLE ASSOCIATION AND "THE EX"

"NRA Convention Vendor Sells Bleeding Female Mannequin Target Called 'the Ex,'" CBS Houston, May 7, 2013.

Erin Durkin and Daniel Beekman, "NRA Blasted for Endorsing Shooting Target That Looks Like Woman and Bleeds," *New York Daily News*, May 7, 2013.

PAGES 254–59: THE FRENCH REVOLUTION, THE REIGN OF TERROR, THE GUILLOTINE, AND MARIE ANTOINETTE

Julia Kavanagh, *Women in France during the Eighteenth Century*, vol. 1 (Smith, Elder, 1850). I quoted "I was a queen, and you took away my crown..." from page 300.

The Kavanagh book is available in another edition: Julia Kavanagh, *Women in France During the Eighteenth Century* (G. P. Putnam's Sons, 1893).

For details about Marie Antoinette bleeding during her trial, see Antonia Fraser, *Marie Antoinette: The Journey* (Weidenfeld and Nicholson, 2001).

For the "I have just been sentenced to death..." quote, see Michael Seed, ed., *Assurance: An Anthology* (Continuum, 2001), 109.

Robert Frederick Opie, *Guillotine: The Timbers of Justice* (Sutton, 2003). See page 70 for the Louis xvi quote, page 146 for the quote about the social isolation of executioners in the Reign of Terror, and page 88 for reference to the man who was beheaded after committing suicide. Additional details come from pages 15, 49–52, and 64–70.

For biographical details and information about Marie Antoinette's health in her last years, see the website *Marie Antoinette*, at http://forum.marie-antoinette.org.

The Louis xvi quote "I hope that my blood may cement..." comes from Jeremy Mercer, *When the Guillotine Fell: The Bloody Beginning and Horrifying End to France's River of Blood, 1791–1977* (St. Martins, 2008), 125.

PAGES 262–64: HARRY POTTER

The "filthy mudblood" quote comes from J.K. Rowling, *Harry Potter and the Chamber of Secrets* (Bloomsbury, 1998).

See "Mudbloods and Murmurs" article on *Wikibooks*, http://en.wikibooks.org.

For details about blood purity in the Harry Potter series, see the "Blood Status" article, http://harrypotter.wikia.com/wiki/Blood_Status; Valerie Frankel's essay "Harry Potter and the Rise of Nazism," at http://frankelassociates.com; and J. K. Rowling's quote about blood issues and the Holocaust, at www.hp-lexicon.org.

PAGES 265–69: THE SPANISH INQUISITION

Erna Paris, *The End of Days: A Story of Tolerance, Tyranny, and the Expulsion of the Jews from Spain* (Lester, 1995).

PAGES 270–73: GENOCIDE

Samuel Totten and William S. Parsons, eds., *Centuries of Genocide: Essays and Eyewitness Accounts,* 4th ed. (Routledge, 2013). From this anthology I drew especially from the following essays: Ben Madley, "The Genocide of California's Yana Indians"; Rouben P. Adalian, "The Armenian Genocide"; Ben Kiernan, "The Cambodian Genocide, 1975–1979"; and Gerald Caplan, "The 1994 Genocide of the Tutsi of Rwanda."

Also see the long and meticulously detailed book by Ben Kiernan, *Blood and Soil: A World History of Genocide and Extermination from Sparta to Darfur* (Yale University Press, 2007).

CHAPTER 5

PAGES 276–79: MY FAMILY HISTORY

I drew the story of my great-great-grandmother Maria Coakley, and her descendants, from various sources:

A telephone interview in June 2013 with my aunt Doris Hill Cochran, who lives in Virginia.

An in-person interview in 2000 with my aunt Jeanne Hill Flateau, who lived in Brooklyn, New York.

Pages 120–22 of my memoir, *Black Berry, Sweet Juice: On Being Black and White in Canada* (HarperCollins, 2001).

Memories of my late father, Daniel Grafton Hill III (1923–2003), who spoke often about Maria Coakley, Marie Coakley, and the challenges faced by my grandparents Daniel and May Hill in the early years of their marriage.

In 2007, I researched and wrote an online exhibit about my father, Daniel Grafton Hill. It is entitled *The Freedom Seeker: The Life and Times of Daniel Grafton Hill*, and available at the Archives of Ontario website: www.archives.gov.on.ca/en/explore/online/dan_hill/index. aspx.

I fictionalized the story of my grandparents, and the efforts by Marie Coakley to break them apart, in my novel *Any Known Blood* (HarperCollins, 1997).

PAGES 283–84: CRIME AND PUNISHMENT

Fyodor Dostoevsky, *Crime and Punishment*, trans. Constance Garnett (P. F. Collier and Son, 1917).

PAGES 287–88: *THE COLOR OF WATER*

James McBride, *The Color of Water: A Black Man's Tribute to His White Mother* (Riverhead Books, 1996). The quote comes from pages 50–51.

PAGE 289: ADOLF HITLER

Adolf Hitler, *Mein Kampf*, trans. Ralph Manheim (Houghton Mifflin, 1971). The quote is drawn from page 286.

PAGES 289–90: EDITH HAHN BEER

Edith Hahn Beer with Susan Dworkin, *The Nazi Officer's Wife: How One Jewish Woman Survived the Holocaust* (Rob Weisbach Books, 1999). Quotes and details come from pages 214–16 and 226.

PAGES 291–92: CHILDREN DURING THE HOLOCAUST

Life in Shadows: Hidden Children and the Holocaust, exhibition at the United States Holocaust Memorial Museum, www.ushmm.org/museum/exhibit/online/hiddenchildren/index.

Simon Jeruchim, *Hidden in France: A Boy's Journey under the Nazi Occupation* (SCB Distributors, 2012).

PAGES 293–94: WAYNE GRADY

Wayne Grady, *Emancipation Day* (Doubleday Canada, 2013).

The quote from Grady comes from an email he sent me on May 14, 2014.

PAGE 295: BELLE DA COSTA GREENE

Heidi Ardizzone, *An Illuminated Life: Bella da Costa Greene's Journey from Prejudice to Privilege* (W. W. Norton, 2007).

Wikipedia, "Belle da Costa Greene."

PAGES 295–300: PASSING, ANATOLE BROYARD, AND PHILIP ROTH

Bliss Broyard, *One Drop* (Little, Brown, 2007). I have taken quotes from pages 10 and 12. Pages 3–17 give a good overview of Anatole Broyard's passing from black into white identity.

Anatole Broyard, *Kafka Was the Rage: A Greenwich Village Memoir* (Carol Southern Books, 1993).

Philip Roth, *The Human Stain* (Vintage, 2001).

For details about fraudulently claiming Blackfoot identity — a fascinating subject but unexplored in this book — see Karina Vernon, "The First Black Prairie Novel: Chief Buffalo Child Long Lance's *Autobiography* and the Repression of Prairie Blackness," *Journal of Canadian Studies* 45, no. 2 (Spring 2011).

PAGES 300–301: JOHN HOWARD GRIFFIN

John Howard Griffin, *Black Like Me* (Signet, 2011).

PAGES 301–2: THE KU KLUX KLAN IN OAKVILLE, ONTARIO, IN 1930

The Ku Klux Klan burned crosses in Oakville, Ontario, on February 28, 1930, to threaten the life of a black man (Ira Johnson) who planned to marry a white woman (Isabella Jones). They married anyway, although Johnson — a World War I veteran who had been born and raised in Oakville's black community — was sufficiently intimidated by the incident to find it necessary to tell the media that he was not actually black but Cherokee.

I have a chapter on this incident in my book *Black Berry, Sweet Juice: On Being Black and White in Canada* (HarperCollins, 2001). See page 222 for the quote from the *Toronto Star*.

I fictionalized this KKK incident in my novel *Any Known Blood* (HarperCollins Canada, 1997).

On the same subject, see Constance Backhouse, *Colour-Coded: A Legal History of Racism in Canada, 1900–1950* (published for the Osgoode Society for Canadian Legal History by University of Toronto Press, 1999). Of particular interest is chapter six, "It Will Be Quite an

Object Lesson: *R. v Phillips and the Ku Klux Klan* in Oakville, Ontario, 1930," which focuses on the charges and court cases stemming from the incident. Only one man was punished: a Hamilton chiropractor by the name of William A. Phillips, who was fined $50 for wearing a mask by night. When Phillips appealed the conviction, the Ontario Court of Appeal sentenced him to three months in jail.

PAGES 303–8: THOMAS JEFFERSON AND SALLY HEMINGS

There are many books and articles about the nearly four-decades-long love affair between American president Thomas Jefferson and his slave mistress, Sally Hemings. I will mention the sources I found most helpful:

Barbara Chase-Riboud, *Sally Hemings* (St. Martin's, 1979). This novel is interwoven with many bits of historical information. See page 262 for the provocative quote from Dolley Madison (wife of James Madison, the fourth American president), and page 341 for the quote from Thomas Jefferson's first draft of the Declaration of Independence.

For a visual representation of Thomas Jefferson's fulminations against King George III in the first draft of the Declaration of Independence, see page 272 of my novel *The Book of Negroes*, illustrated edition (HarperCollins Canada, 2009).

Jan Lewis, "Thomas Jefferson and Sally Hemings Redux: Introduction," *William and Mary Quarterly*, 3rd series, 57, no. 1 (January 2000).

For the excerpt from James Thomson Callender's newspaper article alleging Jefferson's long-time affair with Hemings, see B. R. Burg, "The Rhetoric of Miscegenation: Thomas Jefferson, Sally Hemings, and Their Historians," *Phylon* 47, no. 2 (1986).

For a meditation on the paternity of Sally Hemings's children, based on the timing of visits by Thomas Jefferson to the Monticello plantation where Hemings worked, see Fraser D. Neiman,

"Coincidence or Causal Connection? The Relationship between Thomas Jefferson's Visits to Monticello and Sally Hemings's Conceptions," *William and Mary Quarterly*, 3rd series, 57, no. 1 (January 2000).

For Jefferson's quote on "the amalgamation of whites with blacks," see E. M. Halliday, *Understanding Thomas Jefferson* (Harper Perennial, 2002), 153.

PBS *Frontline* quiz that indicates that Jefferson wrote the quote in 1814. See www.pbs.org/wgbh/pages/frontline/shows/jefferson/quiz/12.html.

The *Wikipedia* article "John Wayles" indicates that Sally Hemings was the half-sister of Jefferson's wife.

Annette Gordon-Reed, *The Hemingses of Monticello: An American Family* (W. W. Norton, 2008).

Wikipedia, "Sally Hemings."

PAGES 308–14: RACE, ANCESTRY, AND GENETICS

Sheldon Krimsky and Kathleen Sloan, *Race and the Genetic Revolution: Science, Myth, and Culture* (Columbia University Press, 2011). From this anthology, I especially drew from the article by Troy Duster, "Ancestry Testing and DNA: Uses, Limits and Caveat Emptor." The quote is drawn from Duster's article, page 113.

Carolyn Abraham, *The Juggler's Children: A Journey into Family, Legend and the Genes That Bind Us* (Random House Canada, 2013). This book provides an up-to-date picture of just how far a person can run with genetics in exploring her family ancestry.

Edward Ball, *The Genetic Strand: Exploring a Family History Through DNA* (Simon and Schuster, 2007).

Henry Louis Gates, *In Search of Our Roots: How 19 Extraordinary African Americans Reclaimed Their Past* (Crown Publishers, 2009). Of particular interest to me were the chapters about Whoopi Goldberg, Oprah Winfrey, and Bliss Broyard.

Troy Duster, *Backdoor to Eugenics*, 2nd ed. (Routledge, 2003).

Katharina Schramm, David Skinner, and Richard Rottenburg, *Identity Politics and the New Genetics: Re/Creating Categories of Difference and Belonging* (Berghahn Books, 2012).

ACKNOWLEDGEMENTS

I WOULD LIKE TO extend first thanks to my wife, Miranda Hill, who supported me in every way imaginable over the course of this book project. Miranda helped me conceive the idea for this book about blood, reassured me that I could do it when my own doubts surfaced, and offered a stream of suggestions about research and revisions. She was my first reader, an astute critic, and a motivating cajoler. Our children — Geneviève, Caroline, and Andrew Hill and Eve and Beatrice Freedman — all pitched in to see me through this project. Geneviève read and commented on numerous drafts, Caroline offered research on vampires and tainted blood, Andrew helped organize my office and files, Eve informed me about Artemisia Gentileschi, and Beatrice asked every month, "How's that book coming?"

I ALSO WISH TO THANK John Fraser, Master of Massey College at the University of Toronto, for finding and funding two diligent, imaginative researchers. Taylor Martin, a mechanical engineer currently enrolled in a graduate program in health administration at the University of Toronto, covered medical and scientific research questions. James McKee, a doctoral candidate in political science at the University of Toronto, researched social, cultural, and historical issues. Both Taylor and James sent me research notes, scholarly articles, dissertations, and book references, always operating with efficiency and grace under pressure. Abbie Buckman, Marilyn Verghis, Caroline Hill, Geneviève Hill, and Miranda Hill also provided valuable research assistance.

MY GRANDFATHER, THE AMERICAN theologian and African Methodist Episcopal Church minister Daniel G. Hill Jr., used to ask me, while he leaned on his cane, to "prop me up on every leaning side." I thought of my grandfather as I relied, all too heavily, on scholars, lawyers, and physicians to prop up my early drafts on every leaning side. They were all generous enough to offer constructive criticism, advice, and encouragement as I waded in waters in which they have swum for years.

Chris Andersen, a Métis scholar in the Faculty of Native Studies and Director of the Rupertsland Centre for Métis Research at the University of Alberta.

Sports journalist Stephen Brunt, on boxing and about performance-enhancing drugs in sport.

Marie Carrière, Associate Professor of French and Comparative Literature and Director of the Canadian

Literature Centre at the University of Alberta.

Avram Denburg, paediatrician and Haematology/ Oncology Fellow at the Hospital for Sick Children in Toronto.

Daniel Heath Justice, a Colorado-born Canadian citizen of the Cherokee Nation and Associate Professor and Chair of the First Nations Studies Program at the University of British Columbia.

Audrey Macklin, Professor and Chair in Human Rights Law at the University of Toronto Faculty of Law.

Minelle Mahtani, Assistant Professor of Geography at the University of Toronto, and author of the forthcoming *Mixed Race Cartographies: Resisting the Romanticization of Multiraciality in Canada* (University of British Columbia Press, 2014).

Émile Martel, a poet who has translated Sor Juana's poetry into French in *Écrits profanes: un choix de textes* (Écrits des Forges, 1996).

Eric M. Meslin, Director, Indiana University Center for Bioethics, and Associate Dean for Bioethics, Indiana University School of Medicine.

Judith H. Newman, Associate Professor of Hebrew Bible/Old Testament and Early Judaism, Emmanuel College and the Department for the Study of Religion, University of Toronto.

André Picard, who has written extensively about Canada's tainted-blood scandal.

Jana Rieger, Professor, Faculty of Rehabilitation Medicine, University of Alberta.

Jean Teillet, Métis lawyer, great-grandniece of Louis Riel, and partner in the law firm Pape Salter Teillet.

Karina Vernon, Assistant Professor of English at the University of Toronto.

Michael K. Schuessler, a professor in the Department of Humanities at the Universidad Autónoma Metropolitana in Mexico City.

Sukanya Pillay, an international lawyer who works in Toronto with the Canadian Civil Liberties Association.

THESE KIND SOULS SET ASIDE TIME for interviews with me:

Bruce Baum, Associate Professor of Political Science at the University of British Columbia, on the historical roots of racist ideology.

Lawyers Margaret Rosling and Thomas R. Berger, Q.C., with the Vancouver law firm Aldridge and Rosling, on Aboriginal identity.

Daniel Coleman, Professor of English at McMaster University, on blood in religious texts.

Glen Coulthard, a member of the Yellowknives Dene First Nation and Assistant Professor of First Nations Studies and Political Science at the University of British Columbia, on Aboriginal identity.

My aunt Doris Hill Cochran, on matters of family culture.

My mother, Donna Hill, on family culture and on the story of Harry Narine-Singh.

Wayne Grady, writer, on racial passing in his family.

Karen Grose, educational leader and superintendent

with the Toronto District School Board, on adoption.

Amanda Jernigan, poet, on blood and identity.

Cangene Corporation employee Cheryl Lawson, on the history of plasma donations in Winnipeg, Manitoba.

Winnipeg resident Raymonde Marius, on donating her own plasma more than one thousand times.

University of British Columbia sociologist Renisa Mawani, on blood and identity.

Ania Szado, writer, on von Willebrand disease.

Trent Stellingwerff, Senior Exercise Physiologist with the Canadian Sport Institute, on the effect of exercise on the bloodstream, and on doping and performance-enhancement drugs in sport.

Patty Solomon, Associate Dean of Rehabilitation Science at McMaster University, on blood and stigma.

Mark Wainberg, Director of the McGill University AIDS Centre and Professor of Medicine at McGill University, and Kris Wells, Associate Director of the Institute for Sexual Minority Studies and Services at the University of Alberta, both on blood donation policies pertaining to gay men.

FOR IDEAS, SUGGESTIONS, CORRECTIONS, contacts, meals, and all manner of encouragement during this writing project, I also wish to thank writers Margaret Atwood, Randy Boyagoda, David Chariandy, Wayson Choy, and Pura López-Colomé; scholars Neil Brooks, Carol Duncan, Christl Verduyn, and Jack Veugelers; educators Grace Centritto, Tina Conlon, and David Cristelli;

lawyers David Cohn, Bryan Finlay, and Seth Weinstein; marathoner Reid Coolsaet and hurdler Perdita Felicien, both Olympians; physicians Anjali Anselm, Mark Crowther, Hertzel Gerstein, and David Price; journalist Lisa Robinson; social worker Robyn Smith; the professors and students at the Universidad del Claustro de Sor Juana in Mexico City; and Pierre Sved, Gloria Antoinetti, Shauna Hemingway, and Ginette Martin at the Canadian embassy in Mexico City; as well as my friends Jennifer Conkie, Carol Finlay, Richard Longley, Jeanie MacFarlane, David Morton, Myra Novogrodsky, Alice Repei, David Steen, Jane Walker, and Gayle Waters.

THANKS AS WELL TO my assistant, Lauren Repei, for her good cheer and diligent work; my agent, Ellen Levine at Trident Media Group, for her dedicated negotiations on my behalf; my devoted, hardworking, and ever encouraging editor, Janie Yoon at House of Anansi Press; Peter Norman for copy-editing; Gillian Watts and Chelsey Catterall for proofreading; Philip Coulter and Bernie Lucht at CBC Radio for assistance with the book and with its adaptation to public lectures and radio broadcasts; and to John Fraser, Anna Luengo, Amela Marin, Liz Hope, and Hannah Allen at Massey College at the University of Toronto, for housing me in Toronto during research trips.

THANK YOU TO the staff at the University of Toronto's Roberts Library. I scoured countless books and articles, and couldn't have done it without you. Sometimes I wonder if there is a book in the world that is *not* in your stacks. Please know that I accepted the invitation to write this book and deliver the 2013 Massey Lectures first and foremost to score a fully loaded University of Toronto library card. Thank you for that. Could I get another, next year?

INDEX

Mubarak, Hosni, 220
murder, 30, 65, 169; as
  "blood on one's hands,"
  282–85; DNA testing
  and, 231–32, 286–87,
  311–12; as entertain-
  ment, 252–53; in films,
  220, 242–43; as
  genocide, 269–73;
  honour killings as,
  80–82; as senseless,
  251–52; of Spanish
  Jews/Muslims, 225,
  265–69, 313; of wives/
  ex-wives, 81–82,
  253–54; by witches/
  demonesses, 221–22,
  224; by women, 82–87
Muruganantham,
  Arunachalam, 42–44
Muslims, Spanish
  persecution of, 225,
  265–69, 313. See also
  Islam
Myrdal, Gunnar, 193

Napoleon Bonaparte, 27
Narine-Singh, Harry,
  186–87
National Association of
  Black Social Workers
  (NABSW), 160–61
National Cancer Act
  (U.S.), 16
National DNA Data Bank
  (RCMP), 231
National Institutes of
  Health (U.S.), 90
National Rifle Association,
  253–54
Native Canadian Centre
  (Toronto), 215
Native peoples: American,
  195–99; Canadian,
  200–1, 203, 205, 206,
  207–12, 213; as defined
  by blood quantum,
  196–99, 208–13;
  genocide of, 273;
  legislation governing,
  195–98, 208–12; as

mixing with other
  "races," 197–99, 200–7,
  212–13; "status"/"non-
  status," 208–13. See also
  First Nations
Nature (magazine), 98, 303
Nehru, Jawaharlal, 150
Neill, Richard, 40–42
Newman, Paul, 218–19
New York City Marathon,
  119
New York Times, 40;
  Broyard's career at,
  295–300
Nietzsche, Friedrich: Thus
  Spoke Zarathustra,
  234–35
Niger: author as volunteer
  worker in, 134–41
Nixon, Richard, 16
Nolen, Stephanie, 42–44

Obama, Barack, 95,
  171–72, 213
Oklahoma City bombing,
  99
Oliver, Percy Lane, 31–32
Olympic Games, 69, 70,
  71, 120; in London,
  67–69, 126, 127; in
  Seoul, 120–24
"one-drop" rule, of black
  identity, 192–93, 213
Ontario Human Rights
  Commission, 59
Opie, Robert Frederick,
  258–59
Örvar Odd (legendary
  warrior), 165–67
Ottenberg, Reuben, 31
Owens, Jesse, 70, 123
oxygen, 8, 21, 214; anemia
  and, 46, 67–69, 126;
  athletic training and,
  72–73, 126–27, 138; red
  blood cells and, 17–18,
  19–20, 21, 132, 138

Paley, William S., 305
Palmater, Pamela, 208–11
Paris, Erna, 267–69

Parmenides, 37
"passing," 287–302; from
  black to Aboriginal,
  301–2; from black to
  white, 156–57,
  293–300; as Holocaust
  survival strategy,
  289–93; from Jewish to
  Christian, 289–93;
  from white to black,
  287–88, 300–1. See also
  entry below
"passing" (specific
  examples): by Broyard,
  295–300; by Grady's
  father, 293–94; by
  Greene, 295; by Griffin,
  300–1; by Hahn Beer,
  289–90, 293; by
  Holocaust children,
  291–92; by Johnson,
  301–2; by McBride's
  Jewish mother, 287–88;
  by Slaney's ancestors,
  156–57
Passover, 11, 63; and "blood
  libel," 265–66
Pasteur, Louis, 24, 50
Patchett, Ann, 88–89
Paz, Octavio: Sor Juana; or,
  The Traps of Faith, 235,
  237, 238–41
Pearl Harbor, Japanese
  attack on, 102, 174
Pelkey, Arthur, 246
performance-enhancing
  drugs, 120–24, 130, 132.
  See also blood doping
Peterson, Merrill, 305
Phelan, Michael L., 212–13
Philostratus, 119
phlebotomy. See
  bloodletting
Pick, Alison: Far to Go,
  155–56
Pierpont Morgan Library
  (New York), 295
Pizzato, Mark, 76–77
plasma, 14–15, 16, 21, 46;
  and blood doping
  scandal, 129, 132; and

(THE CBC MASSEY LECTURES SERIES)